THE KOLLECTED
KODE VICIOUS

THE KOLLECTED KODE VICIOUS

OPINIONATED ADVICE FOR

PROGRAMMERS

George V. Neville-Neil

✦Addison-Wesley

Boston • Columbus • New York • San Francisco • Amsterdam • Cape Town
Dubai • London • Madrid • Milan • Munich • Paris • Montreal • Toronto • Delhi • Mexico City
São Paulo • Sydney • Hong Kong • Seoul • Singapore • Taipei • Tokyo

Many of the designations used by manufacturers and sellers to distinguish their products are claimed as trademarks. Where those designations appear in this book, and the publisher was aware of a trademark claim, the designations have been printed with initial capital letters or in all capitals.

The author and publisher have taken care in the preparation of this book, but make no expressed or implied warranty of any kind and assume no responsibility for errors or omissions. No liability is assumed for incidental or consequential damages in connection with or arising out of the use of the information or programs contained herein.

For information about buying this title in bulk quantities, or for special sales opportunities (which may include electronic versions; custom cover designs; and content particular to your business, training goals, marketing focus, or branding interests), please contact our corporate sales department at corpsales@pearsoned.com or (800) 382-3419.

For government sales inquiries, please contact governmentsales@pearsoned.com.

For questions about sales outside the U.S., please contact intlcs@pearson.com.

Visit us on the Web: informit.com/aw

Library of Congress Control Number: 2020942789

Copyright © 2021 Pearson Education, Inc.

Cover images: Little Princess/Shutterstock, alexslb/Shutterstock, silky/Shutterstock

All rights reserved. This publication is protected by copyright, and permission must be obtained from the publisher prior to any prohibited reproduction, storage in a retrieval system, or transmission in any form or by any means, electronic, mechanical, photocopying, recording, or likewise. For information regarding permissions, request forms and the appropriate contacts within the Pearson Education Global Rights & Permissions Department, please visit www.pearson.com/permissions.

ISBN-13: 978-0-13-678824-9
ISBN-10: 0-13-678824-6

1 2020

To Kaz, who had to hear these often angry outbursts before they ever reached the page.

Contents

Foreword by Donald E. Knuth (*DK*) xi
Preface xiii
Acknowledgments xvi
About the Author xvii

1   The Kode at Hand 1
    1.1   Resource Management 4
    1.2   Big Big Memory 7
    1.3   Coded Arrangements 10
    1.4   Code Abuse 13
    1.5   Nesting Tendencies 16
    1.6   Choking on Changes 18
    1.7   Cursed Code 21
    1.8   Forced Exceptions 24
    1.9   A Nice Piece…of Code 28
    1.10 Something Rotten in… 31
    1.11 Logging 34
    1.12 Lost 37
    1.13 Copying 39
    1.14 Top Five Koding Peeves 42
    1.15 Linguistically Lost 45
    1.16 Check in Comments 48

2   Koding Konundrums 51
    2.1   Ode to the Method 52
    2.2   How Much + in C++? 55
    2.3   Something Sleek and Modern 58
    2.4   What's in a Cache Miss? 61
    2.5   Code Spelunking 65

	2.6 Input Validation 73
	2.7 Dickering with Docs 76
	2.8 What's in the Foo Field? 80
	2.9 Testy Tester 83
	2.10 How to Test 86
	2.11 Leave the Test Modes In! 90
	2.12 Maintenance Mode 93
	2.13 Merge Early 96
	2.14 Multicore Manticore 99
	2.15 This Is Not a Product 102
	2.16 Heisenbugs 105
	2.17 I Don't Want Your Dirty PDFs 109
	2.18 Pining for a PIN 112
	2.19 Reboot 115
	2.20 Code Scanners 117
	2.21 Debugging Hardware 119
	2.22 Sanity vs. Visibility 123
3	**Systems Design** **127**
	3.1 Abstractions 129
	3.2 Driven 133
	3.3 Driven Revisited 136
	3.4 Changative Changes 140
	3.5 Threading the Needle 143
	3.6 Threads Still Unsafe? 146
	3.7 Authentication vs. Encryption 149
	3.8 Authentication Revisited 152
	3.9 Authentication by Example 155
	3.10 Cross-Site Scripting 160
	3.11 Phishing and Infections 166
	3.12 UI Design 172

	3.13 Secure Logging	176
	3.14 Java	181
	3.15 Secure P2P	185
4	**Machine to Machine**	**189**
	4.1 Stepping on Toes	190
	4.2 Paucity of Ports	193
	4.3 Protocol Design	196
	4.4 Which Came First?	200
	4.5 Debugging the Network	203
	4.6 Latency	208
	4.7 Long Distance Runaround	211
	4.8 The Network Is the Computer	215
	4.9 Failure to Scale	219
	4.10 Port Squatting	221
	4.11 Networking in the Raw	224
	4.12 Pointless PKI	227
	4.13 Standard on Standards	230
5	**Human to Human**	**235**
	5.1 Of Pride and…	236
	5.2 What Color Is Your…?	239
	5.3 Broken Builds	242
	5.4 What Is Intelligence?	245
	5.5 Review the Design	248
	5.6 The Naming of Hosts	252
	5.7 Hosting an Interview	256
	5.8 Mythical	260
	5.9 The Obsolete Koder	263
	5.10 With Great Power…	267
	5.11 The Letter	270

5.12 The Tickets That... 273
5.13 Of Screwdrivers and Hammers 276
5.14 Security Reviews 279
5.15 Getting Back to Work 284
5.16 Open Source Licenses 287
5.17 So Many Standards 290
5.18 Books 293
5.19 More on Books 297
5.20 Keeping Up to Date 299
5.21 For My Last Trick 302

Index **305**

Foreword by Donald E. Knuth (*DK*)

Dear DK,

My job keeps me too busy to read real books about computer science. And I don't seem have a great attention span. But I know that everything about computers keeps changing rapidly, and I'm afraid that I'll soon be obsolete if I don't keep up with the field.

Can you suggest any reliable source of current information, by which I might painlessly improve the quality of my work?

Harried Information Hider

Dear Harried,

For many years DK has been a fan of the regular columns by Kode Vicious in Communications of the ACM. The topics are not only timely, they're explained with wit and elegance. KV is not afraid to take unpopular views, and he savagely dissects lots of the insanity that tends to be spreading around.

So DK thinks you ought to try it out. In fact, there's even a better way now, because KV has gathered his columns together and extended them into a book. That book may be just what you crave.

About attention span, on the other hand, that's a tougher problem—especially for people of your generation. Consider spending some time on a desert island, with no access to the internet. Just go somewhere where the weather and accommodations are nice. Take a good technical book with you and lots of scratch paper and pencils and erasers.

A pedagogical book that's full of exercises with worked answers would be especially useful. In fact, if you happen to choose one of DK's own books, you might even find that it contains a quotation by KV himself.

Of course you should also take a copy of KV's book, to keep you grounded.

DK

Dear DK,

Somebody told me that you're a regular reader of Kode Vicious's column, which features answers to letters that he supposedly receives.

When I look closely at those letters, however, it seems to me that they are too perfect. Nobody ever sends me letters that are so well written and to the point.

Do you believe KV forges those letters, or are they actually real?

Skeptical Inquirer

Dear Skeptical,

Indeed, that's exactly the question that DK asked KV, when meeting him in person at the Hackers Conference some years ago. And KV shamelessly admitted to ghost-writing.

But if you think about it, you'll probably agree that the question-and-answer format is an ideal way to express ideas and to teach others. DK even bets that Plato himself ghost-wrote the "dialogues" that Socrates supposedly once had.

Guess what: That format is so effective, DK is now tempted to try it himself.

DK

Preface

What's the worst that could happen?

Famous last words

Welcome to an endeavor I never thought to undertake, the first book of Kode Vicious. In fact, I never thought I'd write a column for a magazine or that that column would run for more than 15 years and more than 100 articles, but life is full of strange twists and turns, especially when you don't duck quickly enough when a table full of your peers is looking for a victim...I mean volunteer!

"So now I'd like to throw out the worst idea of all time." With these words, from Wendy A. Kellogg, the idea that was to become Kode Vicious was born. "It should be someone from the board. Someone with an attitude problem. Someone bald." Back in the early days of *Queue*, I was the only bald board member, though at that point I'd already been shaving my head for a decade.

In February 2004 I was, along with the rest of the *Queue* editorial board, attending our monthly meeting where we get together and try to come up with interesting topics, and authors for *Queue*. It was the early days of the magazine, then in its fourth year, and though we had had several successful issues, we had no regular columnists. I had been invited to the board meetings by Eric Allman, and then written a couple of pieces for the magazine, and was working on co-authoring my first book, but I had never been a columnist, and although the idea seemed fun at the time, perhaps due to too much wine at dinner, I was at a loss as to how to make it actually work.

The original idea for KV, as he came to be known, was actually for a more *Miss Manners* style of column, based on the famous work of Judith Martin, who I had read with my mom when I was a kid. I would write the pieces, *in drag* as it were, and this seemed like an interesting challenge. The first name for KV was *Mother Code,* and I submitted two pieces to our editors based on this persona.

A bit of the character sketch from our meeting might give a better idea of where this was going at the time: Although Mother is never harsh in her advice or criticism, she is also firm in her beliefs. The image is of a strong, but flexible and kind, advice giver. She also has a signature line on every piece, something like "Don't forget to wipe your shoes" or "Remember to wear your galoshes" but that is related to our audience. Something like, "And remember, make sure your code builds before you check it in to the source tree."

In the end this all turned out to be unworkable for a couple of reasons. The most important reason why the original pieces didn't work is that it's very hard to write as someone you're not. Although one or two pieces might have been possible in a very different guise, it's far easier to

xiii

write as someone closer to your own persona than it is to write as someone completely different. Let's face it, Miss Manners I ain't.

I actually spent quite a while trying to come up with the persona I would use, including some obvious ones like "Code Confidential" and "Code Critic" as well as the embarrassing "Captain Safety," "Bug Basher," and "Lint Picker" before hitting on the word Vicious as a good one to use as a *nom de plume*. From there it was a quick romp through "Kid Vicious," "Code Vicious," and "Vicious Kode" to finally getting something that sounded right, "Kode Vicious."

With the new name came a new character sketch:

> @$$hole with a heart of gold. Always willing to teach, but is unwilling to teach those who are not willing to learn. Think Zen monk in a Sex Pistols T-shirt who you worry about bringing to dinner. Often uses nose tweaks to show the student the way or at least *a* way.

I was off and running. I rewrote the original *Mother Code* piece, "So Many Standards," about picking coding standards, and began my career as a columnist.

So, is the author behind Kode Vicious really a big loud jerk who throws co-workers out windows, flattens the tires of the annoying marketing guy, drinks heavily, and beats and berates his co-workers? The answer is both yes and no.

KV is a caricature, and people who know me and have worked with me can easily see how I can write the pieces that I do. Of course, KV is someone I might want to be, or turn into, from time to time, a Hyde to my Dr. Jekyll. Usually I want to be KV when I'm in one of *those* meetings where I take off my glasses, drop them loudly on the table, and run my hand over my bald head, thinking, "How can anyone be so stupid?!" If you are ever in a meeting with me and I do this, it's a tell: Whoever just spoke is a moron. The fact is that beating, or berating, stupid people never makes them any smarter, so instead I turn those thoughts into articles for KV, who can rant about things without winding up in jail and hopefully do a small amount of good at the same time.

It's odd to think about literary influences for KV, but as with any writer, I have several, not the least of whom was my mother, whom I wrote about in "Standards Advice"[1] and who was a hard@$$ and a harsh critic. My favorite authors have always been harsh, direct, and looking to mess with people, and if I'm honest, a lot of KV comes from wishing I was Hunter S. Thompson in the three truly great books he wrote: *Hell's Angels*, *Fear and Loathing in Las Vegas*, and *Fear and Loathing on the Campaign Trail '72*. The more salacious and surreal moments come from reading William S. Burroughs, and if you think it's *Naked Lunch*, I laugh, because that's a nice children's bedtime story compared to *The Wild Boys*.

Another direct influence has been *Queue* itself. Talking to our board and our guest experts for 15 years, the people who come to *Queue* meetings and help us frame our issues, reading and reviewing articles for the magazine, has been one of the most amazing learning experiences of my career. I have been lucky enough to have some truly amazing minds, slightly besotted by wine, violently pointing steak knives across the table at me and telling me just why some idea was either interesting or total bull$#!+.

---

1. https://queue.acm.org/detail.cfm?id=1687192

I have been asked on several occasions, including in one letter, whether I write both the questions and the answers. When I was originally writing the pieces, there were no letters, so I had to write the questions as well as the answers. At first this was quite difficult. I would be staring at the screen, hours away from deadline (I always submit my pieces on or just after the deadline), and I would have nothing besides "Dear KV" in my editor buffer. I then learned a good trick, which I have used ever since. If I run out of material, all I have to do is open up a piece of source code and read it. If the code is well-written, then I can write about what is good about the code, and if, as is often the case, the code is not well written, then I just have to wait for my blood to reach boiling and off I go.

In fact, I use both submitted letters and my own ideas in the column. Whenever something truly stupid happens in a piece of code, during a project, or on the news, I make note of it, if I think I can turn it into an article.

I also get e-mail that is put directly into the question section of the column, and ACM then rewards the letter writer with a *Queue* trinket, which I am sure they will cherish to their dying day. Which pieces are which? I'm not telling, and my editor doesn't know which is which either, so don't try bribing him at the next Turing Dinner.

Of course, KV would not be who he was without his editor, Jim Maurer, who has, for the past 15 years, read KV and then turned these insane ravings into something that ACM would not only publish, but publish in two magazines, *Queue* and *Communications of the ACM* (CACM). From time to time I read a really great line in the finished KV and I'm always intrigued, did I write that or did Jim? I will say that Jim has turned a lot of average bits of writing into something that is actually intelligent and enjoyable, and for this he has my deepest thanks.

I had been planning this book for a while, but it's hard to maintain the level of energy it takes to write as KV for more than about 1,500 words at a time, at least without substances that are only available by prescription, or, since I live in Brooklyn, just down the block. The book finally got written because of the very kinds words of the author of the preface, Don Knuth, who I had asked at one point to write a foreword and who surprised me at a conference with the words, "I've written a wonderful foreword for your book," at which point I knew that I had to finish this work, or the crushing guilt would kill me.

I've kept writing the columns in the meantime, of course, because my writing comes from anger, and anger is what I'm good at. Anger also leads to the dark side, and the dark side has cookies. The book, well, it clearly got done, the question is how long the rehab will take.

George Neville-Neil
aka KV
Brooklyn, NY
June 30, 2020

Acknowledgments

Producing this book was no easy task, and I'd like to thank the following people who turned my nightmare into a much better reality. The crew at Pearson, starting with my editor Debra Williams Cauley who was kind enough to take on this project and who has always willingly listened to my harebrained ideas over cups of coffee for me and tea for her. Julie Nahil, my production editor, and Chris Zahn, my development editor, were those most responsible for turning the chaos I produced into the slick package you're now holding in your hands, or viewing in your e-reader.

Tom Lehrer was kind enough to grant permission for my use of his song "Lobachevsky" as an epigraph for my section on plagiarism.

Eric Allman and Kirk McKusick reviewed this entire book, which means I owe them several good meals at the very least. Their comments and guidance were immeasurably helpful in letting me know when I was on the right track or off in the weeds.

Matt Slaybaugh was instrumental in helping me to find the right pieces for this volume with his research and excellent ACM Digital Library skills.

Jim Maurer has been my editor at ACM *Queue* for the entirety of my tenure as Kode Vicious. He has kept me to a schedule for more than 15 years, no small feat that. When I lived in Japan, I could cheat on my deadlines because I was always a day ahead of him. Jim is the person most responsible for taking my rants and making them interesting and readable to an actual audience. Both I and KV, my alter ego, owe him a great debt for his work all through these years.

My husband, Kaz Senju, has had to live with my Jekyll and Hyde act for even longer than I've been writing as KV. He has graciously put up with both me and the alter ego for nearly 23 years, and without his love, nothing creative I do would be possible.

About the Author

George V. Neville-Neil hacks, writes, teaches, and consults on security, networking, and operating systems. A FreeBSD Foundation board member, since 2004 he has written the "Kode Vicious" column for *Queue* and *Communications of the ACM*. He is a member of ACM's *Queue* Editorial Board; a member of Usenix Association, ACM, and IEEE.

George co-authored *The Design and Implementation of the FreeBSD Operating System, Second Edition* (Addison-Wesley, 2015), with Marshall Kirk McKusick and Robert N. M. Watson. He earned his bachelors degree in computer science at Northeastern University in Boston, Massachusetts. Outside of work in computing and open source projects, George is an avid traveler and speaks several languages, including Japanese, French, Dutch and some Mandarin. He is also an ardent bicyclist. George currently lives in Brooklyn, New York, although he spends about a third of his life traveling for various projects.

1

Welcome to working at the coal face of software.

Anon

The Kode at Hand

Some days you can stare at the same bit of code for hours and make no progress, either in comprehension or extension. When you're looking at files full of code, or even a single line, there are many things that come to mind about how best to craft the code that you're working on, whether it is a new green field or an old mine field handed to you to clear and fix.

Broadly speaking, there are two main areas with which you should be concerned when working on the kode at hand: style and substance. A lot of people have a lot to say about coding styles, and KV has put his oar in in Section 5.17. The most important part of any coding style is to have one. The number of times that KV has looked at code that looked either like an old ransom note or as if every few lines a new koder had mangled it into a shape that they liked is, though countable, maddening to think about. I don't care if you prefer tabs or spaces, just use them consistently! We use our visual systems to understand code, and therefore we must pay attention to how our visual systems use and process information. Poorly laid out code, with randomly chosen variable name conventions, is a recipe for bugs, because our eyes, when they aren't busy trying to look away from the code vomit in front of us, most often miss important points that would have been obvious if the code had a consistent style. At a minimum, a coding style must define a naming scheme and an indentation scheme. The naming scheme defines how function and variable names are chosen. Some languages, such as Go, impose their own requirements in this area, and so the naming scheme needs to be compatible with the language in which you're working. Do you prefer CamelCase? That's fine, just use it throughout. What about noun-verb, such as `fileOpen()` for function names? To some extent it doesn't matter if you use `fileOpen()`, `file_open()`, or `open_file()`, just so long as you use the same system consistently throughout the code. Using `fileOpen()` and `closeFile()` in the same code base is definitely not allowed.

Indentation is probably one of the most common, and stupidest, fights among koders from various backgrounds. Python went so far as to not allow the programmer to choose the indentation; it is dictated by the language. But for most languages we have the two space vs. four space vs. tab (which most often looks like eight spaces) religious war. KV has koded in all three of these systems and personally prefers four spaces because it gives sufficient visual cues without making deeply nested functions go off the right side of his editor window. The only indentation KV hates is the two space, because even with his best glasses it's easy to miss where the indentation of a block of code starts.

Once there is a well-established style, then you can think about the substance of the code, what it actually does, and how it achieves the goals originally set out for it. The substance of a piece of code can be judged in three categories: correctness, conciseness, and composeability.

Of course, a piece of code should be correct (if it's not, you have a bug), but being able to judge a piece of code correct or not is an important part of deciding whether or not you are going in the right direction. Overly long functions, or functions that try to do too many things in one place, make it much harder to judge the correctness of a piece of code. So too does the propensity to make a function for every trivial transformation of data, which leads to hundreds of similarly named functions that are hard to discern from each other.

The preceding discussion mentioned both overly long and overly short functions in code, which brings us to conciseness. A function is concise if it can be fully understood by your future self or another programmer without resorting to writing notes that, in themselves, are longer than the function in question. While I'd not expect to be able to understand every function I look at the moment I looked at it, I do expect that it should take no more than 15 minutes of reading a function to understand its overall intent. A classic counter-example to this concept can be found in the early versions of the `tcp_output()` and `tcp_input()` functions within BSD-derived operating systems. It is a well-known joke that both of these functions are long, torturous, and nearly impossible to understand on first, second, or fiftieth reading. They are so incredibly complex that a months-long project was undertaken just to break them down into something more concise and tractable. It has often been said that, "If it was hard to write, it should be hard to understand," but this is the opposite of our true goal, which is to take complex concepts and break them down such that they can be understood and automated, for automation is the goal of computer science, from whence koding springs. A concise function is one that encodes a single relevant computation or transformation such that it can be understood and, in the best cases, reused, which leads to our last concept, composability.

Large and unwieldy functions are not only hard to prove correct and hard to understand but also nearly impossible to compose into larger systems, which is at the heart of all software engineering. It is certainly possible to build a large system in a single `main()` function, but there are well-established reasons why this is a terrible idea. All systems, and software is a system, are built up from component parts. Making a good component means making something that can work well with other components. Imagine we are building a desk. We could go out into the woods and find a large tree, cut it down, and carve it into the most beautiful desk, lovingly hand-crafted, sanded, sealed, and shellacked. For a number of years the desk serves us well, and we can sit and write and admire the craftsmanship before us, until some part of the desk breaks and needs repair. If we had built the desk out of reusable parts, then it would be simple enough to, for instance, replace a leg that had broken or acquired a wobble, but a desk built out of a single piece of wood, no matter how beautiful at first, is unmaintainable. The heart of good software is being able to adapt, extend, and repair what you have built. Systems change, assumptions change, and if your software is not built out of parts that are easy to compose, then it will not be possible to adapt, extend, or repair it.

This chapter contains letters from disparate and desperate koders who just can't look away from the code on their screen, though they wish that they could. Here we discuss both issues of style and of substance and hope that the readers of KV's responses create systems that have both style and exist in a state of grace.

1.1 Resource Management

> *Data expands to fill the space available for storage.*
>
> Parkinson's Law of Data

One of the perennial arguments in software development has to do with just how much memory is appropriate to use for a particular task. Early koders, working with very limited memory machines, would often compete to see how efficiently they could use the memory on these early computers, where memory was measured in the kilobytes, or sometimes just hundreds of bytes. One of the effects of Moore's law, which made computer memories ever cheaper and more plentiful was to, supposedly, free most programmers from these limited memory constraints. The coming of virtual memory systems and the process model, as well as programming languages that hid how memory was allocated and used, pushed the everyday programmer further from any understanding of the resources they were using in their programs. The elevation of programmers away from the nitty-gritty of memory management has both upsides and downsides. On the upside, programmers can produce more functionality more quickly if we free them from the burden of worrying about whether or not their use of an extra bit of memory here or there will result in their program failing to run. The downside is that all resources are, eventually, finite, and there are real consequences to wasting memory, just like wasting any other resource. No matter how cheap memory is, eventually you're going to run out if you waste it, and the question is then what to do next. If you've never had to contend with limited memory, then you're going to have to learn the same painful lessons of the early koders.

1.1 Resource Management

Dear KV,

I've been reworking a device driver for a high-end, high-performance networking card and I have a resource allocation problem. The devices I'm working with have several network ports, but these are not always in use; in fact, many of our customers use only one of the four available ports. It would greatly simplify the logic in my driver if I could allocate the resources for all the ports no matter how many there are when the device driver is first loaded into the system, instead of dealing with allocation whenever an administrator brings up an interface. I should point out that this device has a good deal of complexity and the resource allocation isn't as simple as a quick malloc of memory and pointer jiggling and a lot of moving parts inside this thing.

We're not talking about a huge amount of memory by modern standards, perhaps a megabyte per port, but it still bothers me to waste memory, or really any resource, if it's not going to get used. I'm old enough to remember 8-bit computers with 64 KB of RAM, and programming those gave me a strong internal incentive never to waste a byte, let alone a megabyte. When is it OK to allocate memory that might never be used, even if this might reduce the complexity of my code?

Fearful of Footprints

Dear Footprints,

The answer to your question is easy. It's sometimes OK to allocate memory that might never be used, and it's sometimes not OK to allocate the same memory. Ah, are those the screams of a programmer without a black-and-white, true-or-false answer to the question that I hear? Delightful!

Software engineering, much to your and my chagrin, is the study of trade offs. Time vs. complexity, expediency vs. quality: these are the choices we deal with every day. It's important for engineers to revisit their assumptions periodically, perhaps every year or two, as the systems we work on change under us quite quickly.

Programmers who are paying attention to the systems they use (and I know that each and every one of my readers is paying attention) have seen these systems change dramatically over the past five years, just as they had the five years before that, and so on, back to the first computers. While processor frequency scaling may have paused for the moment (and we'll see how long that moment lasts), the size of memory has continued to grow. It is not uncommon to see single servers with 64 and 128 GB of RAM, and this explosion of available memory has led to some very poor programming practices.

Blindly wasting resources, such as memory, really is foolish, but in this case it's not an engineering trade-off; it's an example of a programmer who is too far from their machine trying to "just make it work." That's not programming; that's just typing. Software engineers and programmers worth their expensive chairs and high salaries know that they don't want to waste resources, so they try to figure out what the best- and worst-case scenarios are and how they will affect the other possible users of the system. Users in most cases are now just other programs, rather than other people, but we all know what happens to a system when it starts to swap things out of memory onto secondary storage. That's right, your DevOps people call you screaming at 3 a.m. That's fine if you're asleep but not if you're drunk, or maybe it's the other way around. Actually, screaming people are never that much fun, except at a concert.

You mention that this software is for a "high-performance" device, and if by that you mean it goes in a typical 64-bit server-class machine, then no one is really going to notice a megabyte, or four, or even eight. A high-end server-class machine is unlikely to have less than 4 GB of RAM. Even if you allocate 4 MB at system startup time, that's one-tenth of one percent of the available RAM. People writing in Java will suck down far more than that just starting their threads. Are you really going to worry about less than a tenth of a percent of memory?

If you had told me that this driver was for some limited-memory-size embedded device, I would give other advice, since that system might not have 4 GB of RAM; but then again, given what most phones and tablets have in them now, it might.

People are often right when they say, "Waste not, want not," but it's also important to take your moderation in moderation.

KV

1.2 Big Big Memory

> *No one will ever need more than 640K of RAM.*
>
> Attributed to, but denied, by
> Bill Gates

Over the course of the years that I've been Kode Vicious, *typical* RAM sizes in computers have gone from a few megabytes to many gigabytes, and hard disks have gone into the terabytes, and yet we seem never to have enough space for all the data we wish to process. As my Grandma used to say, "Your eyes are bigger than your stomach," though I don't think she was discussing RAM sizes. It seems that no matter how much space we get, we always need more, but, do we?

The only back pressure being applied to this trend may be seen in the world of embedded systems, which, with the advent of the Internet of Terror (IoT) has now become more mainstream. It used to be that the only folks who were working in embedded computing were a small cadre of specialists building systems for planes, trains, and automobiles, and these systems were where we continued to see small memory sizes. Although most systems with the IoT moniker have much more RAM than the more deeply embedded systems of yore, they remain far behind their larger, server-based cousins in data centers, with main memories measured in megabytes, or perhaps a single gigabyte. As cheap compute devices continue to proliferate a new generation of koders must again learn to appreciate the precious resource that is memory and learn to work with it efficiently.

Dear KV,

I've been dealing with a large program written in Java that seems to spend most of its time asking me to restart it because it has run out of memory. I'm not sure if this is an issue in the JVM (Java Virtual Machine) I'm using or in the program itself, but during these frequent restarts, I keep wondering why this program is so incredibly bloated. I would have thought Java's garbage collector would prevent programs from running out of memory, especially when my desktop has quite a lot of it. It seems that 8 GB just isn't enough to handle a modern IDE anymore.

Lack of RAM

Dear Lack,

Eight gigabytes?! Is that all you have? Are you writing me from the desert wasteland where PCs go to die? No one in his or her right mind runs a machine with less than 48 GB in our modern era, at least no one who wants to run certain, very special, pieces of Java code.

While I would love to spend several hundred words bashing Java for, like all languages, it has many sins, the problem you're seeing is probably not related to a bug in the garbage collector. It has to do with bugs in the code you're running, and with a certain, fundamental bug in the human mind. I'll address both of these in turn.

The bug in the code is easy enough to describe. Any computer language that takes the management of memory out of the hands of the programmer and puts it into an automatic garbage-collection system has one fatal flaw: the programmer can easily prevent the garbage collector from doing its work. Any object that continues to have a reference cannot be garbage collected and therefore freed back into the system's memory.

Sloppy programmers who do not free their references cause memory leaks. In systems with many objects (and almost everything in a Java program is an object) a few small leaks can lead to out-of-memory errors quite quickly. These memory leaks are hard to find. Sometimes they reside in the code you, yourself, are working on, but often they reside in libraries that your code depends on. Without access to the library code, the bugs are impossible to fix, and even with access to the source, who wants to spend their lives fixing memory leaks in other people's code? I certainly don't. Moore's law often protects fools and little children from these problems, because while frequency scaling has stopped, memory density continues to increase. Why bother trying to find that small leak in your code when your boss is screaming to ship the next version of whatever it is you're working on? "The system stayed up for a whole day, ship it!"

The second bug is far more pernicious. One thing you didn't ask was, "Why do we have a garbage collector in our system?" The reason we have a garbage collector is because some time in the past, someone well, really, a group of someones wanted to remedy another problem: programmers who couldn't manage their own memory. C++, another object-oriented language, also has lots of objects floating around when its programs execute. In C++, as we all know, objects must be created or destroyed using new and delete. If they're not destroyed, then we have a memory leak. Not only must the programmer manage objects, but in C++, the programmer can also get direct access to the memory that underlies the object, which leads naughty programmers to touch things they ought not to. The C++ runtime doesn't really say, "Bad touch, call an adult," but that is what a segmentation fault really means. Depending on your point of view, garbage collection was promulgated either to free programmers from the tedium of managing memory by hand or to prevent them from doing naughty things.

The problem is that we traded one set of problems for another. Before garbage collection, we would forget to delete an object, or double delete it by mistake; and after garbage collection, we had to manage our references to objects, which, in all honesty, is the exact same problem as forgetting to delete an object. We traded pointers for references and are none the wiser for it.

Longtime readers of KV know that silver bullets never work and that one has to be very careful about protecting programmers from themselves. A side effect of creating a garbage-collected language was that the overhead of having the virtual machine manage memory was too high for many workloads. The performance penalty has led to people building huge Java libraries that do not use garbage collection and in which the objects have to be managed manually, just as they did with languages such as C++. When one of your key features has such high overhead that your own users create huge frameworks that avoid that feature, something has gone terribly wrong.

The situation as it stands is this: with a C++ (or C) program, you're more likely to see segmentation faults and memory-smashing bugs than you are to see out-of-memory errors on a modern system with a lot of RAM. If you're running something written in Java, then you had better pony up the cash for all the memory sticks you can manage because you're going to need them.

KV

1.3 Coded Arrangements

> *We're just rearranging deck chairs on the Titanic.*
>
> Management Adage

Sometimes the problem in a piece of software isn't just the lines of code but how the files that contain the lines are arranged and used. The act of software development is not merely typing in thousands of lines of code, one after the other, but rather it is an act of composition. Well-written software is arranged such that it is easy for others to use; otherwise you might as well make the whole program a single file with a single function and 10,000 or more lines of code. A not unreasonable metaphor is gardening, which we take up next.

Dear KV,

I've been maintaining a set of libraries for my company for the past year. The libraries are used to interface to some special hardware that we sell, and all of the code we sell to our end users runs on top of the libraries, which talk, pretty much directly, to our hardware. The one problem I keep having is that the application programmers continually reach around the library to talk directly to the hardware, and this causes bugs in our systems because the library code maintains state about the hardware. If I make the library stateless, then every library call will have to talk to the hardware, which will slow down the library and all of the code that uses it. What do you think is the right way to get people to actually ask for the features they need instead of reaching around the library?

Tired of the Reach Around

Dear Tired,

I find that the best way to get people to use my libraries correctly is to embarrass them in group meetings by asking questions such as, "What part of 'clean API' design don't you understand?" I usually ask this question in a loud voice and try my best to get the veins in my neck and head to start pulsating wildly. I find that having a generally insane demeanor has the desired effect on my co-workers, as well as bank tellers, people in long checkout lines, and those people who feel they need to stand right next to me on a crowded subway. Another option is to watch every single check-in and to change every program to go through your library, which might be tedious, but don't rule it out. You would not be the first programmer to take that route. In all honesty, unless you enjoy being threatening all day, and I'm told that this is not the normal state for most people, you'll probably have to find out what these so-called application programmers want and try to address their needs.

A good library is like a garden: "As long as the roots are not severed, all is well. And all will be well in the garden. ...In the garden, growth has its seasons. First comes spring and summer, but then we have fall and winter. And then we get spring and summer again." Alas, Chance the Gardener (*Being There*, 1979) didn't really have much to teach anyone about programming or economics, but that's another film altogether. The fact is that we can learn from one particular type of gardener about how to write an effective library.

Consider the following story. A gardener was tasked with placing paving stones so that people could walk through and enjoy the plants and flowers, as well as get to places that were on opposite sides of the garden. A week later the owner of the estate went out into the garden to see how the work was progressing. To his surprise there was not a single stone yet laid anywhere; the garden was pristine but had no paths. Thinking that his gardener must be busy tending to the plants and trees, he decided to ignore what he saw and go back to his office to work. A second week went by and the owner went out into the garden again to see what progress had been made. He was no longer surprised, but more shocked, that the gardener seemed to be continuing to ignore his request for workable garden paths. After a third week went by without any progress, the owner became quite angry and stormed off to find the gardener.

The owner demanded of the gardener, "It has been three weeks! Why haven't you laid a single paving stone? The grass will get torn up by people crossing it any way they please!"

The gardener looked at his boss with apparent shock and said, "But, Sir, how can I know where to put the paths until the people have shown me where they would walk?" He then asked the owner to follow him down to the garden. When they arrived he showed the owner how, rather than walking any which way on the grass, paths had already been worn where the largest number of people needed to cross to their destinations. He turned to the owner and said, "Now that I can see the paths, I can lay the stones."

If you want people to use your library, then you need to find out why they're reaching around it and try to address their needs. The library isn't there for you; it is there for them.

KV

1.4 Code Abuse

> *It can be better to copy a little code than to pull in a big library for one function. Dependency hygiene trumps code reuse.*
>
> — Rob Pike

The line between code reuse, which we tend to laud, and abuse, which we tend to deride, can be a fine one. One koder's macro is another koder's library, or some such nonsense. Given the malleability of software, it is sometimes tempting to reuse a piece of code that really is not fit to the purpose we try to fit it to, and that's code abuse, not reuse.

Dear KV,

During some recent downtime at work, I've been cleaning up a set of libraries, removing dead code, updating documentation blocks, and fixing minor bugs that have been annoying but not critical. This bit of code spelunking has revealed how some of the libraries have been not only used, but also abused. The fact that everyone and their sister use the timing library for just about any event they can think of isn't so bad, as it is a library that's meant to call out to code periodically (although some of the events seem as if they don't need to be events at all). It was when I realized that some programmers were using our socket classes to store strings just because the classes happen to have a bit of variable storage attached, and some of them are globally visible throughout the system, that I nearly lost my lunch. We do have string classes that could easily be used, but instead these programmers just abused whatever was at hand. Why?

Abused API

Dear Abused,

One of the ways in which software is not part of the real world is that it is far more malleable, as you've just discovered. Although you can use a screw as a nail by driving it with a hammer, you would be hard-pressed to use a plate as a fork. Our ability to take software and transmogrify it into shapes that were definitely not intended by the original author is both a blessing and a curse.

Now I know you said you clearly documented the proper use of the API you wrote, but documentation warnings are like yellow caution tape to New York jaywalkers. Unless there is an actual flaming moat between them and where they want to go, they're going to walk there, with barely a pause to duck under the tape.

Give programmers a hook or an API, and you know they're going to abuse it. They're clever folks and have a fairly positive opinion of themselves, deservedly or not. The APIs that get abused the most are the ones that are most general, such as those used to allocate and free memory or objects, and, in particular, APIs that allow for the arbitrary pipelining of data through chunks of code.

Systems that are meant to transform data in a pipeline are simply begging to be abused, because they are so often written in incredibly generic ways that present themselves to the programmer as a simple set of building blocks. Now, you may say these were written as building blocks for networking code, or terminal I/O, or disk transactions; but no matter what you meant when you wrote them, if they're general enough and you leave them in a dark place where other coders can find them, then the next time you look at them, they may have been used in ways unrecognizable to you. What's even better is when people abuse your code and then demand that you make it work the way they want. I love that, I really do...no, I don't!!!

One example is the handling of hardware terminal I/O in various Unix systems. Terminal I/O systems handle the complexities inherent in various hardware terminals. For those too young to have ever used a physical terminal, it was a single-purpose device hooked to a mainframe or minicomputer that allowed you to access the system. It was often just a 12-inch-diagonal screen, with 80 characters by 24 lines, and a keyboard. There was no windowing interface. Terminal programs such as xterm, kterm, and Terminal are simply a software implementation of a hardware terminal, usually patterned on the Digital Equipment Corporation's VT100.

Back when hardware terminals were common, each manufacturer would add its own special, sometimes very special, control sequences that could be used to get at features such as cursor control, inverse video, and other modes that existed on only one specific model. To make some sense of all the chaos wrought by the various terminal vendors, the major variants of Unix such as BSD and System V created terminal-handling subsystems. These subsystems could take raw input from the terminal and, by introducing layers of software that understood the vagaries of the terminal implementations, transmogrify the I/O data such that programs could be written generically to, say, move the cursor to the upper left of the screen. The operation would be carried out faithfully on whatever hardware the user happened to be using at that moment.

In the case of System V, though, the same system wound up being used to implement the TCP/IP protocol stack. At first glance this makes some sense, since, after all, networking can easily be understood as a set of modules that take data in, modify it in some way, and then pass it to another layer to be changed again. You wind up with a module for the Ethernet, then one for IP, and then one for TCP, and then you hand the data to the user. The problem is that terminals are slow and networks are fast. The overhead of passing messages between modules isn't significant when the data rate is 9600 bps; but when it's 10 Mbps or higher, suddenly that overhead matters a great deal. The overhead involved in passing data between modules in this way is one of the reasons that System V STREAMS is little known or used today.

When the time finally came to rip out all these terminal I/O processing frameworks (few, if any, hardware terminals remain in service), the number of things they had been extended to do became fully apparent. There were things that were implemented using the terminal I/O systems pretty much as a way to get data into and out of the operating-system kernel, completely unrelated to any form of actual terminal connection.

The reason these systems could be so easily abused was that they were written to be easily extended, and one programmer's extension is another programmer's abuse.

KV

1.5 Nesting Tendencies

> *It's turtles, all the way down.*
>
> From Hindu mythology

Among the many frustrations of koding, trying to find the definition of a structure or variable that is hidden deep within a branching forest of `include` files ranks quite near the top. Even with code spelunking tools (see Section 2.5), this task is often nontrivial. Perhaps we ought to think more carefully about how deeply or broadly we wish to build our include tree, before we make all such searches resemble digital needles in haystacks.

Dear KV,

A co-worker recently chewed me out for including one C file within another C file. Though this may not be the way many people build software, it doesn't seem wrong to do so, and the compiler doesn't complain. Is this really wrong, or just unorthodox?

All in One and One in All

Dear All in One,

There are many things that a compiler will not complain about. Compilers, unlike people, have no moral sense and therefore have only a very limited idea of right and wrong. Most people have a better sense of right and wrong than a compiler—most, but not all. Clearly, you are one of those people who does not have a highly developed moral sense, for if you did, you would have realized, even before e-mailing me, that what you were doing was wrong. Why was it wrong?

The simplest reason not to include one C file within another is that it obfuscates the relationships between pieces of code. The few times I've seen this kind of thing in practice, it has been done at the end of a C file, sort of like tacking on one file at the end of another. If, for example, you don't wind up reading the very end of the file, you'll never know that there is one file (or more) built when the code you're looking at is compiled. Such surprises are an unwelcome part of a programmer's day. It's like grabbing a bag that you thought contained a few loaves of bread only to find out that there is an anvil at the bottom.

If you need to use a piece of code in two places, you don't #include it in both; you build it as a separate module and use the linker to put the pieces together at the end when you build the final executable.

We are no longer programming in the 1950s, and we are not building programs out of paper-tape libraries where you splice together a program out of smaller segments of paper tape. In an environment with linkers, loaders, and compilers capable of generating and working with separate modules, there is no excuse for #including one compilable file within another.

KV

1.6 Choking on Changes

Make everything as simple as possible, but not simpler.

Albert Einstein

The following piece was written in the years before distributed version control systems, such as git, were in common use. Now, in my daily koding life I think of this piece often, usually as often as when I merge most other koders' branches into my own. The ability to go off and work privately on a branch that will eventually be merged to the main line has only exacerbated the tendency of some koders to make massive sets of unrelated changes and to then push all of these as one big hunk, which must be painfully swallowed in the next merge. Although all the guidance on working with a DVCS is to the contrary, there remains a stubborn minority of koders who simply do not get this idea, that changes should be as small and easy to understand, by the writer and by the eventual reader, as possible.

Dear KV,

I recently got a dressing-down by my boss because I made a large change to our system in one large chunk. I didn't think it was too large, only about 20 files and a couple of thousand lines of code, but I was forced to back out my change and then commit it in smaller bits to our repository. While I understand that people don't like large changes because they can be potentially destabilizing, all of this code was interrelated and really couldn't be intelligently broken up into smaller pieces of work. In your experience, what's the optimal size for a single check-in?

Lot o' Lines

Dear Lot,

First, I rather doubt that you got a dressing-down by your boss. I am sure HR would frown on undressing engineers, or anyone else, in the workplace.

More seriously, the question of how large a check-in should be allowed to a source base is not one to which there is a hard and fast answer. Some features, or bug fixes, require widespread changes to a code base.

I think many less experienced engineers and many managers suffer under the false belief that bug fixes usually require a small number of lines to be changed and that features require a large number of lines. If most bugs were caused by one error or other hard-to-find but easy-to-fix one-liners, then that would be true, but it's not. Often a bug will be pervasive, infecting an entire code base. A pervasive bug is usually the result of a false assumption being coded into an entire system, with the result that the fix touches many, if not most, files. When such a fix is required, it makes sense to check in all the changes in one fell swoop. It hardly makes any sense not to fix the same bug, at the same time, in all places.

Where many programmers run into trouble is when they have run-on code. Run-on code is like a run-on sentence. You're working on a feature, but then you find a bug and then you realize that there is another related bug and you fix the second bug, which leads you to a third problem and you fix that as well, all the while your project lead is asking for the feature that you started implementing when you found the first bug, so you go back to feature development but haven't checked in any of your fixes because you need to have the feature complete to test the fixes, and at the end you're left, like our readers, out of breath and unable to remember where the whole thing started.

Part of being an engineer is taking large systems, or problems, and breaking them down into chunks that are small enough for you and the majority of your co-conspirators, I mean colleagues, to understand and digest. No one wants to read 2,000 lines of code spread over 20 files in one shot. It probably took you more than a week to develop that feature and fix those bugs, and you should not expect people to be able to come up to speed on what you've done in only a few minutes by reading your, I am sure, lengthy expository commit message.

One final reason to break up large commits is so that if someone finds a bug in some part of what you've written, it might be possible to roll back a smaller change and preserve the rest of the work that you've done. The alternative, which is to commit a large chunk of work and then patch, patch, patch, is sloppy and unsatisfying.

My guidelines here are simple. If you change an API, then change it and all its consumers at once. If you fix three bugs and add a feature, then you need to make four distinct check-ins.

KV

1.7 Cursed Code

> *Perhaps you shouldn't speak in your talks like you do to your friends.*
>
> Advice received after my first public talk at university where I used many words that were OK on a Bronx street, but not in the academy.

Koders often believe that what they write in code is a form of private communication between themselves and the machine, but this belief is mostly a false assumption. Other koders will come after you and look not just at the code but at the comments and other things you leave behind.

There are many things that sound fine inside our heads, which will not look fine when written down, and it is important to learn the difference early on in our koding careers. Of course we have all been in that frustrated state when looking at our work, or the work of others, and we've wanted to let loose with a barrage of colorful metaphors, but these are best screamed aloud at a wall rather than preserved forever in a source code control system. I can only imagine that one of the reasons some companies, Microsoft most famous among them, provide all developers with hard wall offices is to give them something to throw invective, silly putty, and mice against.

As the point of software development is to produce an artifact that is shared among many people, koding should be thought of more as a public performance than a private matter, and therefore we ought to consider, carefully, what we say, and how we say it, in code, in comments, and elsewhere.

Dear KV,

One of the people on my current project keeps complaining about my use of "colorful metaphors" in code. While I understand that I shouldn't be checking these sorts of things into our source repo, I can't see why he complains when he sees them on my screen. I mostly use such words for debugging messages because they're shocking enough to stand out from the rest of the log messages produced by our software. I can't really believe that KV would object to a programmer adding a bit of salt to log messages.

Kolorful Koder

Dear Kolorful,

I can understand why you might think that I might be a prolific user of colorful metaphors, given some of the things I write about in this column. You are correct, and my co-workers can tell you that, because of my occasional outbursts when dealing with particularly horrific bits of code, they have learned a thing or two about body parts and functions that they wish they could now forget.

Unfortunately, for you at least, I have to come down on the side of your co-worker in this dispute. While I'm sure you have faithfully marked every place that your code might exclaim a colorful metaphor with the well-known comment, "XXX Remove This!" the fact is that if you do this enough, someday, and usually on quite the wrong day, you're going to forget. You probably think you won't, but the risk is not worth the eventual hassle. I've been through that hassle, and I'm glad that, for once, the problem wasn't my fault.

More than a decade ago I worked for a company that produced a software IDE (integrated development environment) and some associated low-level software. One of the IDE's limitations on a certain platform was that every project saved by the IDE had to have an appropriate extension: those letters after the dot that provide a clue about what type of file has just been saved. While programmers are quite used to giving their files such descriptive monikers as notes.txt, main.c, and stdlib.h, it turns out that not everyone is familiar with this sort of naming standard, and some even prefer names such as Project1 and Project2, without any type of identifying extension.

The programmer working on the IDE decided that if the user of his program declined to add an extension to the project filename, he would add one for them. He chose a four-letter word that rhymes with duck. I'm not sure if he meant this to go out in the release, as a way of pointing out customers who refused to use file extensions, or if it was something he meant to change before the release, but in the end it didn't matter. Within days of our 1.0.1 maintenance release of the IDE, there was a 1.0.1b release with a single change. I don't remember if the b release had a note saying what changed, but all of the engineers working on the software knew the real reason.

Amazingly, the programmer who did this got to keep his job. I suspect there were two reasons for this, the first being that he was actually a pretty good programmer, and the second being that he was the only one in the company willing to support the IDE on the platform that he was working on.

While this is a pretty extreme example of a colorful metaphor gone wrong, and while I know that there are programmers who will leave extremely strong language in comments, I have to say that I frown on this as well.

Your code is your legacy, and while your mother might never see it, you should still only check in code that would not shock her should she choose to read it.

KV

1.8 Forced Exceptions

> *Programmers are often angry because they're often scared.*
>
> Paul Ford

One might well argue that the software industry, and indeed most technically focused industries, is rife with hubris and pedantry. Sometimes that hubris and pedantry gets translated into our code, most often into the code that forms the basis of the languages and tools we use to create software. Pedantry can be put to good uses and bad, but it's when we have unexamined pedantry that we most often get into problems.

Dear KV,

I subscribe to "The Morning Paper," a daily summary prepared by one person, Adrian Colyer, who curates research papers and sends them out to interested readers (`https://blog.acolyer.org`).

Last fall he reviewed "Simple Testing Can Prevent Most Critical Failures: An Analysis of Production Failures in Distributed Data-Intensive Systems" (`https://blog.acolyer.org/2016/10/06/simple-testing-can-prevent-most-critical-failures/`). It had some surprising results, including:

- Almost all catastrophic failures (48 in total, or 92 percent) are the result of incorrect handling of nonfatal errors explicitly signaled in software.
- Error handlers with TODO or FIXME in the comments. This example took down a 4,000-node production cluster.
- Error handlers that catch an abstract exception type (e.g., Exception or Throwable in Java) and then take drastic action such as aborting the system. This example brought down a whole HDFS (Hadoop Distributed File System) cluster.

And the list went on from there.

I've been reading KV for quite a while, and as I read the review and then the paper itself, it looked like something you would be interested in, so I've sent along the link.

Helpfully Not in Error

Dear Helpfully,

Yes, KV also reads "The Morning Paper," although he has to admit that he does not read everything that arrives in his inbox from that list. Of course, the paper you mention piqued my interest, and one of the things you don't point out is that it's actually a study of distributed systems failures. Now, how can we make programming harder? I know! Let's take a problem on a single system and distribute it. Someday I would like to see a paper that tells us if problems in distributed systems increase along with the number of nodes or the number of interconnections. Being an optimist, I can only imagine that it's $N(N+1)/2$, or worse.

I don't think you pointed out this paper to KV just to hear me bang my head on my desk while thinking distributed systems, so let's assume you're asking the "Why?" question: "Why is it the case that 92 percent of the catastrophic failures in this paper are caused by a failure to handle nonfatal errors?"

Well, let's see what else the paper had to say and then think about how software is actually implemented in the real world, rather than how we believe it ought to be implemented in the illusory world that management and marketing inhabit.

To get to the heart of why nonfatal errors might have led to fatal errors, we need look no further than this snippet from the paper: "This difference is likely because (i) the Java compiler forces developers to catch all the checked exceptions; and (ii) a variety of errors are expected to occur in large distributed systems, and the developers program more defensively. However, we found they were often simply sloppy in handling these errors" (https://www.usenix.org/system/files/conference/osdi14/osdi14-paper-yuan.pdf).

Hopefully anyone who has been a professional programmer for more than a few days knows that many developers will always write the code they are most interested in, or pressured to deliver first, which is not the error- and exception-handling code, nor is it test code, nor documentation, the latter two of which I have already harangued readers about, ad nauseam. What management and the rest of the team want is "the code," and what most people see as "the code" is only the part of it that explicitly does the job you're expected to do. It's not even the demands of others that cause this narrow focus; it's often just that the error-handling parts are not as interesting to the person writing the code as getting a result. It would seem that many programmers just want to move those bits, munge that data, and show pictures of cats.

In point of fact we have a clear indication of the importance programmers put on the error-handling components of the code by this finding: "Error handlers with TODO or FIXME in the comment." Personally, I prefer XXX, as it reminds me of my time in Amsterdam in the early 1990s, and unless you're working in certain industries that might also serve photos, and might still serve photos of cats, you're unlikely to find XXX as a variable in the code.

We can look at the fact that the Java compiler forces programmers to catch all the unchecked exceptions in one of two ways. If we are charitable, and KV is the heart and soul of charity, we assume that the Java language and compiler developers are simply helping programmers make fewer mistakes and make sure that their code not only does what it is meant to do, but also acts appropriately when things go awry.

If we are less charitable, or perhaps more honest and realistic, we see this enforcement quite differently: as a naked attempt to control programmers and make them do what the language and compiler people thought was right at the time. "Programmers don't do proper error handling. I know, we will MAKE them handle errors, or their programs won't compile at all!" I believe this is said in the voice of an overbearing school teacher. "You will dot your i's! You will catch all exceptions!" Except that unlike dotting an i, there are ways to skate around handling the exception that was meant to be handled. In a rush? Well then, just add a TODO or FIXME or XXX in the comments and move on. You'll come back to it later...of course you will.

Both sides are a little bit wrong in this case. We can all point fingers at the person who leaves a trail of FIXMEs in the code, but who among us is without blame in that regard? We can also blame the pedants who thought that forcing every exception to be caught was doing us a favor. You can never discount the human element in programming. For everything you try to force on someone, there is something they will work to avoid if at all possible. Tool builders need to understand that the people who use their tools are often trying to get a very narrow job completed with a minimum amount of effort. Was it wrong to add the forced exception handling into the tool? Maybe and maybe not. In the hands of someone with the time and inclination to do the right thing, these errors are a welcome way of finding problems that they do have to handle.

Clearly, in the hands of a large percentage of programmers who work on some of the most complex systems yet devised, the feature is actually a nuisance, and it is likely time to rethink how this particular exception ought to be handled.

KV

1.9 A Nice Piece...of Code

> *Good programming is 99% sweat and 1% coffee.*
>
> Anon

So much of my time is spent looking at code that, at best, could be considered diamonds in the rough and, at worst, a steaming pile of moose turd pie. Over the years I've tried to bring a few of the better pieces to light and share them as examples of what I might find good and right in software development. What is good and bad in software development could, and does, fill many volumes in many libraries, and much of the advice you'll see winds up being in conflict with itself and others. I have a few, simpler, rules to follow that I think are brought out here. Code needs to be readable by people other than the original author, broken down into small enough pieces to be easily composed into different forms over time, testable, and documented. It seems a simple recipe, so why is it so hard for people to follow?

Dear Readers,

One of the things that can make me really like a piece of code other than the obvious ones such as decent documentation, proper indenting, and rational variable names is when a function or subsystem is properly reused. Over the past month, I've been reading the IPFW (IP Firewall) code (written by Luigi Rizzo at the University of Pisa), which is one of the firewalls available in FreeBSD. Like any firewall, IPFW needs to examine packets and then decide to drop, modify, or pass the packets unchanged through the system. Having reviewed several pieces of software that do similar work, I have to say that IPFW does the best job of reusing the code around it. Here are two examples.

Part of the job of a firewall is to classify packets and then decide what to do with them. There are a few ways to go about this, but what IPFW does is quite elegant. It reuses a tried and tested idea from another place in the kernel, the BPF (Berkeley Packet Filter). The BPF classifies packets using a set of opcodes sort of like a machine language for processing network packet headers to decide whether a packet matches a filter that has been specified by the user. Using opcodes and a state machine for packet classification leads to a flexible and compact implementation of the packet classifier, compared with hand-coding rules for later use. IPFW extends the set of opcodes that can be used for classifying packets, but the idea is exactly the same, and the resulting code is easy to read and understand and therefore easier to maintain and less likely to contain bugs that might let malicious packets through. The entire state machine that executes any and all firewall rules in IPFW is only 1,200 lines of C code, including comments. Another advantage of using a set of opcodes to express the packet-processing rules is that the entire chunk of C code, which is really a bytecode interpreter, can be replaced by just-in-time compiled code, generated by an optimizing compiler. This leads to an even greater increase in packet-processing speed.

A more direct case of reuse is how IPFW directly reuses the kernel's routing-table code to store its own address lookup tables. Many of the rules in a firewall make reference to the source or destination address of a packet. While it is quite possible to write your own routines for storing and retrieving network addresses, and many people have, there is no need to rewrite this code, in particular when your firewall code will already be linked into a program that has such routines available. The radix code in the kernel can manage any type of key/value lookup, although it is optimized to handle network addresses and associated masks. The IPFW table-management code is really just a simple wrapper around the radix code, as can be seen in the following lookup code:

```c
int ipfw_lookup_table (struct ip_fw_chain * ch,
                      uint16_t tbl,
                      in_addr_t addr, uint32_t * val)
{
  struct radix_node_head * rnh;
  struct table_entry * ent;
  struct sockaddr_in sa;
```

```
    if (tbl >= IPFW_TABLES_MAX)
    return (0);
    rnh = ch->tables[tbl];
    KEY_LEN(sa) = 8;
    sa . sin_addr . s_addr = addr;
    ent = (struct table_entry *)
                (rnh->rnh_lookup(&sa, NULL, rnh));
    if (ent != NULL) {
      * val = ent->value;
      return (1);
    }
    return (0);
}
```

All this code does is take arguments understood by IPFW, such as the chain of rules (ch), address table (tbl), and address being sought (addr), and pack them up in a way that is usable by the radix code, which is called on line 13. The value is returned in the last argument to the function. All of the other functions in the table-management code, which add, delete, and list entries in the table, look very much the same. They are wrappers around the radix code. Treating the routing-table code as a library, as IPFW does, means writing less complex and tedious code and results in a mere 200 lines of C code, including comments, to implement tables of network addresses. It is this sort of reuse, not the tortured kind that I more often come across, that leads me to praise this code.

Don't worry, I'm sure next time I'll be back to ranting about some bad bit of code, but I have to say it has been a nice surprise to have found two well-written pieces of code in two months. I think it's some sort of record.

KV

1.10 Something Rotten in...

> *Something is rotten in the state of your code.*
>
> ——————————————
> Apologies to Shakespeare

I admit that kode does not actually have a literal smell, but as the six senses go, smell seems the one most appropriate to some pieces of it. Opening some files is like opening a durian, except that when you eat durian, it tastes great, but that smell, you'll never forget it. Some code literally makes your eyes water, your stomach turn, and you feel as if you need to run from the room for fresh air.

The following letter shows quite the opposite, a piece of code that is clear, concise, well written, and in short does not stink. Let's take a look.

Dear Readers,

Every once in a while, I come across a piece of good code and like to take a moment to recognize this fact, if only to keep my blood pressure low before my yearly medical checkup.

The first such piece of code to catch my eye was clocksource.h in Linux. Linux interfaces with hardware clocks, such as the crystal on a motherboard, through a set of structures that are put together like a set of Russian dolls.

At the very center is the cyclecounter, a very simple abstraction that returns the current counter from an underlying piece of hardware. A cyclecounter knows nothing about the current time, time zone, or anything else; it knows only what the register in a piece of hardware is when asked about it. The cyclecounter has two pieces of state that help translate cycles into nanoseconds but nothing else. The next doll out is the timecounter. A timecounter contains a cyclecounter and raises the level of abstraction to the level of monotonically increasing time, measured in nanoseconds. On top of these structures are others that eventually give the system enough abstraction to know what the time of day is.

So, what is so great about this code? Well, two things: first, it is well structured, in that it is built from small components that can cooperate without giving each other a reach-around or a layering violation; second, it is written and documented in a style that is clear and clean enough that I was able on first reading to understand how it worked.

The comments and structure for the cyclecounter give you a flavor of what makes me so happy to read this code:

```
/**
 * struct cyclecounter - hardware abstraction for a free
 *    running counter.  Provides completely state-free
 *    accessors to the underlying hardware.
 *    Depending on which hardware it reads, the cyclecounter
 *    may wrap around quickly. Locking rules (if necessary) have
 *    to be defined by the implementor and user of specific
 *    instances of this API.
 *
 * @read:    returns the current cycle value
 * @mask:    bitmask for two's complement
 *              subtraction of non 64 bit counters,
 *              see CLOCKSOURCE_MASK() helper macro
 * @mult:    cycle to nanosecond multiplier
 * @shift:   cycle to nanosecond divisor (power of two)
 */
struct cyclecounter {
  cycle_t (*read) (const struct cyclecounter * cc);
  cycle_t mask;
  u32 mult;
  u32 shift;
};
```

I think you can see why I like this code, but just in case you can't, let me be more specific. The code is well laid out, is nicely indented, and has variables that are short yet readable. There are no Bouncy Caps or very_long_names_that_read_like_sentences. The comment is long enough to describe not just what the structure is, but how it is used, and it even mentions what has to be done if multiple threads need to access one of these structures simultaneously. If only all code were this well documented! You can read more of this code online at `https://github.com/torvalds/linux/blob/master/include/linux/clocksource.h`.

My other example of good code requires more explanation, so I'm going to reserve it for a future column. After all, I don't want to waste the calming effect all in the same issue.

KV

1.11 Logging

We're looking for a needle in a haystack.

The start of every debugging session.

I strongly suspect that most koders do not appreciate the importance of logging systems or how logging output is used, and this piece brings out the importance of both. Nearly all debugging of failed systems starts with the log output, if you're lucky enough to have it. While a debugger can tell you where a program crashed, the log output shows what happened in the moments before a crash. Not just limited to helping with crashes, a good logging system can be the basis of a good system dashboard. A bad logging system, and there are plenty of these, is often worse than having no logging at all, because poor logging output often sends you down the wrong path, wasting your time in trying to track down issues in your system.

Dear KV,

I've been revising the logging output for a large project, and it seems every time I propose a change, our systems admins start screaming at me to revert what I've done. They seem to think that the format of our log output was set in stone at version 1 of the product and that I shouldn't actually touch anything, even though the product now at version 3 does quite a bit more than it did in version 1. I understand that changing the output means they'll have to change some scripts, but I can't help it if there are new features that need to log new information.

Log Rolled

Dear Logged,

I don't know if you know this, but systems administrators are simply lazy, drunken layabouts who spend all their days slacking off work, putting their feet up the desk, and sipping single malts while the boss isn't looking. Actually, in point of fact, systems administrators are often the busiest and most harried people at any IT site, and they are the people responsible for knowing if all the systems are UP or DOWN. If you capriciously change the logging output on their systems, the tools they have lovingly crafted to track the performance of your system will indicate things are DOWN when they're probably not, and this will result in a lot of screaming. I like coding in a quiet environment; I do not like screaming, so do not make the sysadmins scream.

There are good ways and bad ways to update log output. Inserting a new column at the beginning of each line, thereby throwing off all the following columns, is an incredibly bad way of updating a log file. In fact, the first columns of any log output should always be the date and time with seconds. Using the date for the first column makes writing analysis scripts far easier. Just like extending a programming API, unless you have a very good reason, you should always add new information at the end of the line. Extra columns are the easiest to ignore and the least likely to cause the sysadmin tools to go nutty. That therefore reduces the amount and volume of screaming in the office (see above about offices and quiet).

Another, less offensive way of updating log output is to add whole new lines of information so that scripts can look at the old lines correctly and, for as long as possible, ignore the new information. Allowing the script authors some time to update their scripts is a kindness that is repaid in free liquor at conferences and is just the kind of thing you would want to encourage.

Finally, you might simply add an option to the program to output the old log format so that the people running your software have time, again, to update their scripts, or perhaps they really don't need the new information and would like the chance to use your system without touching their pristine and beautiful scripts. Think first before forcing new information on the user.

KV

1.12 Lost

> *Your keys and wallet are on the dresser, where they always are.*
>
> Exasperated spouse

You know that feeling you have when you lose something, that maddening feeling that you get when you absolutely *know* that you left that something somewhere but just can't remember where? Well, it turns out that people do this all the time with their source code as well, even after 40-plus years of centralized version control systems. It's another instance of, "It doesn't have to be this way."

Dear KV,

I have this problem. I can never seem to find bits of code I know I wrote. This isn't so much work code that's on our source server, but you know, those bits of test code I wrote last month; I can never find them. How do you deal with this?

Lost

Dear Lost,

Losing things makes me feel stupid, and I hate feeling stupid. Several years ago a bright but very spacey friend of mine pointed out an obvious answer to keeping track of all those things I have on my computers. Put them in a source repository! Not the work one, dummy! The last thing you want is to give your employer all your personal code, documents, and stuff. I mean, what happens when they go under, as they so often do?

A few years ago I took one of my servers at home (you do have servers at home, yes?) and put a CVS repository on it. I now keep everything I care about, which is just about everything, checked into my own personal CVS. This includes all of my configuration files (dot files since I'm a Unix head), all documents (including this letter and response), all code, all scripts, pretty much everything. Anything I create on a computer that I'm going to use more than once goes into the repository. Given the huge amounts of disk space available cheaply nowadays, it doesn't make any sense to throw away anything you think you might use again.

As a side benefit I can now set up my environment on any computer in about ten minutes. I just make sure CVS is on the new machine and check out my Personal/DotFiles directory. Then I run a small script that links all the necessary files from Personal/DotFiles to their correct places and, bang, instant KV koding machine. This can work equally well on non-Unix platforms, but I don't use those, so you'll have to work out the mechanics yourself if you're not on Unix or a Unix-like system.

Oh, and remember to back the repository up in some way. Mirror it, burn it to CD-ROM or DVD, but do something for that "just in case moment." If you think you feel dumb when you've lost one file, think of how absolutely, amazingly stupid you'll feel when you lose a few megabytes, or more.

I'm sure the first letter I'll get after this column will be whining about how to find things once you've been storing everything. I recommend you use *directories*, a feature of operating systems since the 1970s. I also suggest you use reasonable names that do *not* include "foo," "bar," and "baz"!

KV

1.13 Copying

> *Plagiarize!*
> *Let no one else's work evade your eyes.*
> *Remember why the good Lord made your eyes.*
> *So don't shade your eyes, but plagiarize, plagiarize, plagiarize.*
> *Only be sure to please call it, "research."*
>
> "Lobachevsky," Tom Lehrer

Not all of KV's work is call and response; sometimes I take the time to just go off on something that I consider to be important. The ability of computers, and in particular computers connected to the Internet, to allow people to copy and paste code and information has resulted in an explosion of unreviewed and unexamined code. If the unexamined life is not worth living, then the unexamined code is not worth executing. The following advice is also applicable to nearly all things that are the product of intellect, not just code, but documentation, security and other policies, and test plans; pretty much anything that someone produced and wrote down cannot be swallowed whole without re-evaluation.

One of the topics not covered in the following piece is proper attribution. If someone performs a copying operation whole, as described, attribution isn't a problem because the original author's name and other information will be preserved as long as the copier doesn't change anything in the resulting version. A problem arises when people copy code, or documents, and try to make them their own by just slapping their name or their company name all over the document. I have heard people who should know better argue that this sort of plagiarism is both common and a "best practice," but what it actually is is intellectually dishonest. What makes such acts even more ludicrous is that the same search technology that allowed them to find the code or documents can be used by anyone to show the provenance of the plagiarized version. Professors routinely run their students' work through search engines and other systems to find cases of plagiarism; the same should be done with all code and documents both to determine the intellectual provenance of systems and to protect against future bugs and security issues. If you don't know where your code or ideas came from, you're doing yourself and those who consume your works a serious disservice.

Dear Readers,

You know, sometimes I really don't need to read my mail to get going on a subject; sometimes I just need to read a little code. I'm sorry I can't share the code with you, as it is proprietary, something I was looking at for a client, but it brings up two more items in the long list of things that Kode Vicious really hates. The more I think about it, the more I realize it is just one big problem with a lot of different faces. The problem? Computers make it too easy to copy data. Yes, that's right, I know you were all expecting me to rail on about the poor quality of comments, or documentation, but in reality it's just that computers are too good at something they're designed to do. I guess I really shouldn't blame the computers, I ought to blame the idiots behind the keyboards, but it's much more acceptable to take out your anger on machines than on people. After all, you're probably not going to be arrested for chucking a computer out a window, unless it hits someone on the street below, but you can bet that chucking a co-worker out a window, though it might feel good at the time, will have consequences. In order to shed some light on what I am ranting about, let me tell you a story from my day.

The day started out fine, the birds were singing, the sun was shining, all was... Oh never mind that! Today I was looking over a fix a programmer had made to some code that handled C++ strings and C char* buffers. The code had a bug where the string, when placed in the buffer, was not properly terminated, leading to some data leakage into other parts of the system. All well and good, pointers and strings are difficult beasties and the known source of many program errors. So, loaded with my usual level of mid-day caffeine, that is to say, just short of grinding my teeth to dust, I decided to check the fixes. I opened up my editor to one of the files that had been changed, and though the code itself offended me, I was only here to look and did not want to get mired in reworking the fix. Having checked the first change, I went on to the next, which looked like a carbon copy of the first one. OK, well, that's fine, a couple of places isn't a problem. I moved to the next difference, and it too was the same as the two previous. I think you can see how this goes. I went through more than ten changes. The same bug had appeared in more than ten places, in a program with only about 200 files, and why? Because the person who had written the first version of the buggy code had simply copied their bug, over and over again. The code that I'm talking about wasn't just a single-line call to some function; this was 15 lines of code that was handling a known dangerous quantity, a pointer to a buffer, a frequent source of errors.

So, now we come to the first annoyance, the ability to copy and paste code all over the place. I don't want to say, "Never cut and paste code!" because such strong statements don't take enough situations into account, but I can say, "Before you cut and paste code, THINK!" You see, many years ago, long before you, or I, or many people reading this were born, some very nice people invented the function call and, in 1951, the library. It would seem that most people think of libraries as being provided by others but not by themselves. As you all know, a function call is a way of simplifying repetitive work. Instead of doing the same thing with 10 or 20 or, and—yes, I have seen this—100 lines of code copied and pasted all over your software, you simply say, "Oh, look, this code is used again and again; I bet it's generally useful." Then you take that code, put it in a function, put that function in a library, and then you share it with all your co-workers who can thereby benefit from your genius. Just like Mom said when we were kids, "Sharing is good!"

Now, unfortunately, the story doesn't end here. It is on days like today when I actually feel sorry for the people who sit near me, because although I have learned to control my use of extremely vulgar language in the office, people still find the sound of my head hitting my desk to be a disturbing sound. It was that sound that brought the usual calls of, "What now?" from my neighbors, who are occasionally amused by my rantings, as long as I stop hitting my head on the desk.

What I had found, completely by accident, was a whole subdirectory of the current product, which contained a subset of the files from the product I was checking. It would seem that in order to make a new product someone had just copied the old one and started editing it. Now, there weren't just the ten-plus bugs in the code I was supposed to check, but the same bugs in the code that someone had copied to make their new product. But wait! There's more! The new product hadn't retained all of the old code; no, it had kept many of the APIs but had subtly change their underlying meanings, adding a few new constants here, changing a return value there. In a single file, the one that had led me to this dubious discovery, there were 200-plus separate changes, not significant enough that if someone linked with the wrong library they would get an obvious error, oh no—only enough that the code would break in weird and mysterious ways.

So, now we have two different problems caused by the ease of copying. The first problem is the replication of bugs in dangerous, pointer handling code that must now be maintained in ten-plus places in one product. The second problem is the copying of a whole product, all bugs included, and two products that are related but subtly different enough that fixing a bug in one requires manually fixing a bug in another.

All I can wonder is "What were these people thinking?" I mean, yes, for those of us on Unix-like systems it is very easy to type `cp -r OldProduct NewProduct` and then get right to work. Look at all the time we've saved! We are sure to get a raise for our productivity, instead of a kick in the teeth, which is what we deserve. For those with a desktop metaphor in mind, it's just a point and a click, even less work than typing the 28 characters above, including Enter, required for those of us on Unix. That, though, is not the point; the point is that software is made up of libraries and functions for a reason, and when you come across something that you need, you should attempt to make it easily reusable by you and by others. It will save you more time in the long run than you saved by blindly copying code or files. If you work with me, it may also save you from defenestration. Yes, that's the word for today, defenestration, look it up!

KV

1.14 Top Five Koding Peeves

*I've got a little list,
of society offenders,
who might well be underground,
they never would be missed,
they never would be missed.*

The Lord High Executioner
The Mikado

In all the years of writing KV I've only ever resorted to lists for a few pieces. When I went back to include this piece, I was surpised that it was just five peeves, because I knew that even back when this was written, I had to have more than just those. In the intervening years none of these peeves has gone away; they remain at the top of the list, and no technology has ameliorated them in kode that I come across daily.

Dear Readers,

In this issue nasty old Uncle Vicious is not going to print and respond to a letter. Why is that? Well, it's not because y'all have been silent, by any means. This month I want to talk about my Top Five Coding Peeves. And why do I want to talk about this? Well, because I'm tired of seeing them in your code when I come to work where you've been working, that's why! So, in ascending order, here are the things that cause me to slug down unhealthy amounts of alcohol after work.

5) Crappy Comments

You, yes, you, you know who you are. You are the people who write comments like:

```
// Set i equal to 1
i = 1;
```

when that's obvious and then leave huge, complex functions that do nine different things under twenty different cases completely uncommented.

Oh, and how about remembering CS 101 and placing a useful comment at the head of each function? Just because you don't think anyone will see that function doesn't mean someone like me isn't going to have to clean up that mess later.

4) Dangling `else` Clauses

One of the most common errors made when fixing or extending software is the addition of code inside what you thought was a block, which turns out not to be. This is what I call a dangling `else`, and I tend to just add braces when I see them. Why? Well, it's not like the two extra brace characters per else are going to overload your disk, unless you're programming on some very creaky hardware. In that case, ask your mother for $50 to buy a bigger hard disk or delete some of those movies you've been downloading.

3) Magic Numbers

I don't know how many times I've seen this:

```
name_buf[128];  // Hold the name of the file.
```

followed either by `name\_buf` being incorrectly indexed or the very same type of storage declared differently:

```
new_name_buf[127];    // Hold the name of the file.
```

No matter what language you use, C, Java, etc., at some point you're going to need to use constants, and when you do, you should name them. Why? Well, not only is it a cleaner way to code, but it also provides much more consistent results since you're more likely to use the constant name everywhere you need it. Now, what if you port the code to a system where names are shorter or longer? Do you want to search all the code to find these magic numbers and change them?! I didn't think so.

This particular use of magic numbers is lamer than usual as there is a well-known, operating system–defined constant that you should be using. On the machine on which I'm writing this column it's actually 1,024 bytes, and it's called FILENAME_MAX.

2) Code Dingleberries

Nothing is more jarring when reading code than to come across a huge chunk of commented-out code in the middle of a file or function. Since you're using a source code control system (wait, you are using a source code control system, right?), you can simply remove the code and check in the new version. If you ever need the dead code to be resurrected, you can simply retrieve the older version.

The same complaint also goes for blocks of code that are conditionally compiled out at build time using things like `#if`/`#endif`. There are some systems that depend on `#ifdef` for configuration, which I also find problematic, but this is not the same as the `#if 0` case. If you don't want the code to be in there, then don't put it in there, or make your source code control system remember that it was, once there. The code should always represent the system as it is, not as it was, or should be.

1) Global Variables

There is, I sincerely hope, a special level in hell reserved for those programmers who, every time they fail to make a proper abstraction, add another global variable to their programs. On top of this, I hope there is yet another level, with even nastier punishments, for those who name such global variables things like `s` and `r`. Tracking down problems in software written like this is not impossible, but it's not my idea of fun. So please, if you really need a global variable, give it a name that is easy to find with `grep`, or other tools like cscope, and make sure you really, really, really need it.

So that's it, my top five pet peeves. Sorry, no prizes, not even a booby prize, just me screaming, "Stop doing that!" I'll throw the floor open now and let you write to me with your pet peeves. If I like them, I might publish them. If I don't like them, well, there's always the D key.

KV

1.15 Linguistically Lost

What's behind door number one?

"Let's Make a Deal"

Choosing a language in which to write a particular piece of code is a nontrivial exercise, and it requires both patience and experience not only to choose the right language for the job but, often, to convince others that your choice is the correct one. The following piece was written prior to the most recent explosion of new languages, such as Go, Rust, Lua, and others that are now trying to push their way past C++, Java, and Python. Time will tell which of the new languages have staying power, and which will fade, but the advice given in this letter more than a decade ago still stands; you just have to change the names to protect the guilty.

Dear KV,

Where I work we use a mixture of C++ code, Python, and shell scripts in our product. I'm always having a hard time trying to figure out when it's appropriate to use which for a certain job. Do you only code in assembler and C or is this a problem for you as well?

Linguistically Lost

Dear LL,

First of all, I may be a grotty old koder, but I'm not so limited as to only program in C and/or assembler. If you ever suggest that again, I'll beat you senseless with this 6502 manual.

Now, choosing a language to use is not an easy task. The majority of coders just use what they're given at work and don't question that. Not questioning is bad. We should always be trying to find the right tool for the job. Then, once we've found the right tool, we can spend endless hours yelling at our co-workers for not using the same tool we do. No, wait, that's not it.

Let's start from the end of your list. Shell scripts, though sometimes clever and useful, are very difficult to maintain. Uncle Vicious never uses them for anything more than 100 lines of code. Why? Well, I've seen products built on directories filled with shell scripts, and they invariably turn out to be a tangled mess. I'm sure there is someone out there who has the most beautifully built, finely tuned set of shell scripts, and they will all send me nasty hate mail, which, thankfully, will be electronic and easily deleted. The fact is that the majority of people who write shell scripts write them as one-off (or two-off because they never work the first time) things to "just get something to work." Unfortunately, these scripts are the red-headed stepchildren of projects. They are checked in and ignored until they cause a problem. Then they're beaten soundly until the screams die down and they begin to limp along again. `while 1 do echo "previous rant" done`

Python, one of the modern interpreted languages, along with Perl, Ruby, and others, is an appropriate tool with which to replace shell scripts. These too have their limitations, most of which have to do with performance. Please, do not write device drivers, embedded systems, flight control software, or other code that gets between me and the brakes of my car in Python. I tend to prototype things in Python, and then if I decide I need the speed, I move on to a compiled language. I'm sure that several people will write in to say that I should use Java, but I find that Java requires too much other cruft (libraries, IDEs, etc.) for me to use it in the way that I use Python. Someday I might write a large system in Java, just like someday I might write a large system in Modula-3, Smalltalk, or Scheme. This is not what I use Python for. Python is for replacing hundreds of twisty shell scripts, all alike.

Finally, we come to C++. Now, whatever people may say about C++ with respect to other compiled languages, the fact is that it is used by a majority of koders in their day jobs. I prefer C++ for large, complex systems that are pretty static once they're in place. The fact is that when you're at work, it is unlikely that you'll be able to say, "I don't think we should use C++ for this project," unless you're the CTO, and given your question, I rather doubt you're the CTO.

I think one of the things you're really asking is when to put something into C++ vs. Python. Well, if you're having to write a class for general use in this system and others, then you probably want to code it in C++. Furthermore, if the code in question should not be modified or seen by the customer, then C++ is, again, your natural choice. I'm pretty sure your boss doesn't want you sending code snippets of the company's secret sauce out with the product in easy-to-read chunks. The performance question is also important. Is the cost of starting up an interpreter and then interpreting the code going to outweigh benefits of writing in an interpreted language? If so, then you're going to have to write that code in C++.

Brian W. Kernighan and Rob Pike actually did a good job of addressing issues like this in their book *The Practice of Programming*, which is a KV must-read.

So, there are no hard and fast rules here, of course, but this is how I happen to handle these issues in my own work.

KV

1.16 Check in Comments

> *Say what you mean, and mean what you say.*
>
> Anon

And so finally we come to the point where the code at hand is all ready to go into your version control system. You've gotten to this point, the code is ready to be mixed with the rest of the code base, you should be proud of your achievement, you should tell the whole world, or at least you should make a reasonable attempt to describe the change in your commit log.

Dear Kode Vicious,

Regarding your Coding Peeve 2, code dingleberries (Queue, Nov. 2004, p. 20), I agree it's better to clean out unused code. I often comment out areas replaced by revised code because I might want to remember how I did something if I need to do it elsewhere later. Putting it in a source code control system is a good idea, but how can you remember which version it was in years later?

Short-Term Memory

Dear STM,

I'm not sure what source code control system you're using, but all the ones I've used in the last ten years have the ability to store a comment with each check-in. This check-in comment, once stored, can also be retrieved. When I am looking at a piece of code that has faded from my memory, for instance one that was checked in on a Friday before a particularly good party, I will get the entire history of that file and scan it for telltale signs of what I or others were thinking when various changes were made. I'm sure, STM, that you put in proper comments when you check in your code, but I'm also sure that there are plenty of readers who could benefit from a short rant on another of my pet peeves.

Lame check-in comments are a pet peeve that didn't make my top five, but they are certainly in my top ten. Many years ago I actually worked with an engineer who refused to put in any comments when he checked in his code. It was completely in keeping with his manner as he also rarely spoke more than ten words in a day at work and answered all questions with a "yes" or a "no." This lack of communication had nothing to do with his language abilities, as he was fully fluent in English, the language we all used at work. Eventually the team installed a trigger that required at least three words to be put into any comment before a file could be checked in. If the comment was empty, then the check-in was rejected. To be honest, this didn't help us very much as we then got comments like:

Added new feature. Fixed bug 511. Fixed bug 432.

etc.

Although I considered, and suggested, thumb screws be applied to this particular engineer, I was told that that was not a modern business practice, at least in the United States. After a couple of years of this madness, the engineer left the company. Although he was a pretty good koder, there were bugs in his software, just like any of us have. The major difference was that fixing anything that this guy had worked on was a nightmare because none of the code had any context. There were incredibly lame comments in the code, an already-addressed peeve, and the check-in trail told you nothing about the evolution of the code. Working on this guy's stuff was akin to taking your turn in the barrel. All of us did it, but none of us really enjoyed it.

Check-in comments come under the same restrictions as comments in code. In the first place, there should actually be a comment. As far as I'm concerned, all source code control systems should force an engineer to include a comment in a check-in. If it were possible to deliver an electric shock to the keyboard of those who check in code without comments, I would install that too. Second, the comments have to be useful. The three-word stricture placed on the engineer I worked with clearly was insufficient. I figure if you fix a trivial bug, then you should include a full sentence about what the bug number was, so it can be looked up later what the problem turned out to be and how you fixed it. Brevity may be the soul of wit, but unless you are Oscar Wilde, and he's dead, so you're clearly not, you will probably need at least three full sentences to meet my criteria. Of course, when you add something more significant, such as a new feature, then you really need an entire paragraph to describe what the feature is, how it works, and how it is used.

I'm sure that I will get complaints that there are better places to store such information. The bug information should be in the bug database, and the feature information should be in the specification. These are both true, but when I open a piece of code, I want the greatest amount of information available to me in one place. I do not want to have to switch between the bug database window and the electronic version of the spec and my code. I want it all in one place, and I want it now. So, please, next time any of you check in something to your source code control system, pretend that I'm standing behind you with a cattle prod, and consider very carefully what you're about to write.

KV

2

> *The time has come*
> *The Walrus said*
> *To speak of many things*
>
> "The Walrus and the Carpenter,"
> Lewis Carroll

Koding Konundrums

Stepping away from the kode in front of us, we come to a slightly broader set of koding koncepts. Many people fail to understand that programming and software design aren't just about typing hundreds of lines into an editor or IDE and then pressing Run. There are concepts we must address no matter how large or small the system is that we're working with. As any system is built there are the problems of debugging, documenting, and testing the system as well as understanding challenges to the overall system performance, and these are some of the problems we turn to in this chapter.

2.1 Ode to the Method

> *The good thing about science is that it's true whether or not you believe in it.*
>
> Neil deGrasse Tyson

If computer science is truly a science, then clearly the scientific method can be applied to solve problems that present themselves in computers and their accompanying software. A topic rarely covered in computer science programs is how one might actually apply the scientific method to software engineering. If one is not solving a problem via something akin to the scientific method, then can they be said to be debugging by faith? Faith-based debugging has, thankfully, not taken off in the way agile and scrum have, but it is definitely a method I have seen applied by many people who really ought to know better. Treating software bugs as near supernatural occurrences is a frequent joke among koders, "Did you sacrifice a chicken?" being a common retort to someone saying that they simply cannot find a bug or get the printer to work. While there are koders who are so good that they can find the bug just by scanning a page of your kode, these are few and far between, and so the best programmers learn to apply what, in essence, is the scientific method to solving their problems. In the following response I lay out a very simple way to apply the method to fixing software bugs and being confident in the fix.

Dear KV,

I just started working for a new project lead who has an extremely annoying habit. Whenever I fix a bug and check in the fix to our code repo, she asks, "How do you know this is fixed?" or something like that, questioning every change I make to the system. It's as if she doesn't trust me to do my job. I always update our tests when I fix a bug, and that should be enough, don't you think? What does she want, a formal proof of correctness?

I Know Because I Know

Dear I Know,

Working on software is more than just knowing in your gut that the code is correct. In actuality, no part of working on software should be based on gut feelings, because, after all, software is supposed be a part of computer science, and science demands proof.

One of the problems I have with the current crop of bug-tracking systems (and trust me, this is only one of the problems I have with them) is that they don't do a good job of tracking the work you've done to fix a bug. Most bug trackers have many states a bug can go through: new, open, analyzed, fixed, resolved, closed, etc., but that's only part of the story of fixing a bug, or doing anything else with a program of any size.

A program is an expression of some sort of system that you, or a team, are implementing by writing it down as code. Because it's a system, you have to have some way of reasoning about that system. Many people will now leap up and yell, "Type systems!" and "Proofs!" and other things about which most working programmers have no idea and which they are not likely ever to come into contact with. There is, however, a simpler way of approaching this problem that does not depend on a fancy or esoteric programming language: use the scientific method.

When you approach a problem, you ought to do it in a way that mirrors the scientific method. You probably have an idea of what the problem is. Write that down as your theory. A theory explains some observable facts about the system. Based on your theory, you develop one or more hypotheses about the problem. A hypothesis is a testable idea for solving the problem. The nice thing about a hypothesis is that it is either true or false, which works well with our Boolean programmer brains: either/or, black or white, true or false, no "fifty shades of gray."

The key here is to write all of this down. When I was young, I never wrote things down because I thought I could keep them all in my head. But that was nonsense; I couldn't keep them all in my head, and I didn't know the ones I'd forgotten until my boss at the time asked me a question I couldn't answer. Few things suck as much as knowing that you've got a dumb look on your face in response to a question about something you're working on.

Eventually I developed a system of note taking that allowed me to make this a bit easier. When I have a theory about a problem, I create a note titled THEORY, and write down my idea. Under this, I write up all my tests (which I call TEST because, like any good programmer, I don't want to keep typing HYPOTHESIS). The note-taking system I currently use is Org mode in Emacs, which lets you create sequences that can be tied to hot keys, allowing you to change labels quickly. For bugs, I have labels called BUG, ANALYZED, PATCHED, —, and FIXED, while for hypotheses I have either PROVEN or DISPROVEN.

I always keep both the proven and disproven hypotheses. Why do I keep both? Because that way I know what I tried and what worked and what failed. This proves to be invaluable when you have a boss with OCD, or, as they like to be called, "detail oriented." By keeping both your successes and failures, you can always go back, say in three months when the code breaks in a disturbingly similar way to the bug you closed, and look at what you tested last time. Maybe one of those hypotheses will prove to be useful, or maybe they'll just remind you of the dumb things you tried, so you don't waste time trying them again. Whatever the case, you should store them, backed up, in some version-controlled way. Mine are in my personal source-code repo. You have your own repo, right? Right?!

KV

2.2 How Much + in C++?

> *My name is Ozymandias, King of Kings*
> *Look on my Works, ye Mighty, and*
> *despair!*
>
> —"Ozymandias,"
> Percy Bysshe Shelley

The language wars are never ending, and in the following letter KV was perhaps baited into bashing one such language. To be honest I've always hated C++; I find it bloated and that it mostly produces intractable mounds of horse$#!+, as my grandmother used to say, which are hard to maintain and hard to tune. My response here is probably a bit tamer than my true feelings on C++, in part because I know that people will choose what they choose and also that we're all dealing with a mound of technical debt that we often didn't start with but that was thrust upon us on day one at work. That there are schools that still think that C++ is a good language to start teaching computer science with is, quite frankly, maddening, and I wish they would stop. I consider that particular practice to be mental health abuse. It would be better to give the students Python than C++, or, perhaps, an assembler.

In the years since this letter and response were first published we find that there are at least two new contenders in the compiled language space, Rust and Go. KV has only a passing acquaintance with both of these but is excited to see Rust trying to push into the embedded space, in particular. A language with better memory safety in systems that often lack virtual memory protections seems like it ought to be a good thing for every koder's mental health. The other upside with both Rust and Go is that they look somewhat familiar to those of us who already program in the Algol-like languages, so a jump from, say, C, C++, or Java to either of these languages isn't going to cause so much cognitive dissonance that you feel as if you'd put a small piece of blotter paper under your tongue before starting to work. As for either of these new languages as a teaching language, the jury is very definitely still out, and I still prefer we teach students with something interactive to start, such as Python.

For now let's see just how much + KV was able to find in C++.

Hello KV,

Since there was some debate in my company over this issue, I'm curious to see what do you believe: putting aside performance issues (which I think are relatively minor on modern PCs), when would you recommend using C++ for development, and when would you recommend C? Do you think it is always better to use C++?

My feeling is that unless your application is inherently object-oriented (e.g., user interfaces), C++ will tend to make the implementation worse instead of making it better (e.g., constructors and operators doing funny unexpected things, C++ experts trying to "use their expertise" and writing C++ code that is very efficient but extremely hard to read and even not portable, huge portability and performance issues when using templates, incomprehensible compiler/linker error messages, etc., etc.). I also think that while people can write bad C code (gotos out of macros was a nice one), typically people can write *awful* C++ code.

So, what do you think, where do you stand on this dispute?

Wondering How Much + There Is in ++

Dear Wondering,

Picking a language is something I've addressed before in other letters, but the C vs. C++ debate has raged as long as the two languages have been in existence, and really, it's getting a bit tiring. I mean, we all know that assembler is the language that all red-blooded programmers write in! Oh, no wait, that's not it.

I'm glad you ask this question, though, because it gives me license to rant about it and also to dispel a few myths.

The first and most obvious myth in your letter is that user interfaces are inherently object-oriented. While many introductory textbooks on object-oriented programming have user interfaces as their examples, this has a lot more to do with the fact that humans like pretty pictures. It is far easier to make a point graphically than with text. I have worked on object-oriented device drivers, which are about as far as you'll ever get from a user interface.

Another myth that your letter could promulgate is that C is not an object-oriented language. A good example of object-oriented software in C is the vnode filesystem interface in BSD Unix and other operating systems. So, if you want to write a piece of object-oriented software, you can certainly do it in C or C++, or assembler for that matter.

One final myth, which was actually dispelled by Donn Seely in "How Not To Write FORTRAN in any Language" (ACM Queue vol. 2, no. 9 - Dec/Jan 2004–2005), is that C++ leads to less understandable code than C. Over the past 20 years I have seen C code that was spaghetti and C++ code that was a joy to work on, and vice versa.

So, after all that myth bashing, what are we left with? Well, the things that are truly important in picking a language are:

1) What language is the largest percentage of the team experienced in?

    If you're working with a team and six out of eight of them are well-versed in C but only two know C++, then you're putting your project, and job, at risk in picking C++. Perhaps the two C++ koders can teach the C folks enough C++ to be effective, but it's unlikely. In order to estimate the amount of work necessary for a task you have to understand your tools. If you don't normally use a nail gun, then you're likely to take someone's toe off with it. Losing toes is bad as you need them for balance.

2) Does the application require any of the features of the language you're using?

    C and C++ are a lot alike as languages, i.e., in syntax, but they have different libraries of functions and different ways of working that may or may not be relevant to your application. Often real-time constraints require the use of C because of the control that can be gotten over the data types. If type safety is of paramount importance, then C++ is a better choice because that is a native part of the language that is not present in C.

3) Does the application require services from other applications or libraries that are hard to use or debug from one or the other language?

    Creating shim layers between your code and the libraries you depend on is just another way of adding useless, and probably buggy, code to your system. Shim layers should be avoided like in-laws. They're OK to talk about, and you might consider keeping them around for a week, but after that, out they go as so much excess, noisy baggage.

There are lots of other reasons to choose one language over another, but I suspect that the three listed should be enough for you and your team to come to some agreement, and you'll notice that none of them had to do with how easy it is to understand templates or how hard it is to debug with exceptions.

KV

2.3 Something Sleek and Modern

Move fast and break things.

Idiots' Mantra

Many letters to KV run along this vein of wanting to use a new language or technology to kode. As I point out here, what matters isn't newness but appropriateness to the job, and the use of good software engineering practices. Most of us in the koding business have a thing for the sleek, new, and modern. Many in our industry would probably be quite comfortable with the Futurist Manifesto, except for its unhealthy relationship to fascism. Written in 1909 by Filippo Tommaso Marinetti, nearly the entire manifesto could be twisted into the language of modern tech companies, "Move fast and break things," but honestly the manifesto was twisted enough to begin with.

As technologists many of us are attracted to new ideas, methods, and ways of working, especially if they promise to make our systems faster or able to do more with less. The fact is that being sleek and modern is no panacea for the ills of software. Good software requires careful thought, deliberate planning, and careful execution, as I point out here, and there has yet to be a silver bullet to the problems of developing good software.

Dear KV,

When I read your column, you sound to me like one of these guys who only kodes, as you would misspell it, in C or maybe C++. Many of us code in other languages such as PHP, Python, and Perl. How about writing about languages like those? Perhaps something written and designed after most of your readers born.

Where I work we provide a lot of web services, so we use a good deal of PHP in our work, with only a small amount of C and C++ doing the computationally intensive tasks, or things that have to be closer to the operating system. Do you have any advice for those of us who kode in other languages?

21st Century Kodern

Dear 21,

I'll have you know that I was born after C, but only just. I have no doubt that many of the readers of Queue cut their teeth on other languages, perhaps, though I shudder to think it BASIC, but I can only write about what people ask about. There have been letters from koders using languages other than C or C++, such as Linguistically Lost, who wrote to me a few months ago. If you had asked a concrete question, this would have made my job easier, but clearly that was not your goal. As to my misspellings, take it up with my editor, though I recommend you take a couple of self-defense classes first.

I have read plenty of PHP code in my time as well as Python, Perl, C, C++, TCL, Fortran, Lisp, COBOL, and others. The basic fact is that what separates good code from bad code has very little to do with the language itself. As someone recently pointed out in Queue, you should learn "How Not to Write FORTRAN in Any Language."

Good code is code that uses the dominant metaphors in a language in order to make it easy for other people to understand. All the languages I've mentioned in this reply have comments, and yet many people seem to either leave these out or misuse them completely. Since the 1980s it has been possible to write understandable variable and function names, and yet people still continue to use single letters, believing that those who look at the code after them will know what they mean.

In PHP you can write this:

```
function getn($data)
```

as easily as this:

```
// This function takes a string as input in the name_field
// The name must be a string starting with an alphabetic
// character, i.e. A..Z or a..z, and may not be more than 32
// characters in length. Only the first character must be
// alphabetic and all the following characters can be
// alphabetic or numeric i.e. 0..9
function get_name($name_field)
```

and yet people continue to write the first version. So, I don't care how modern you and your young friends are; if you apply the basics of writing code that is easy to understand the first time, some other poor koder has to read it. Once you're done with that, get back to me with some specific questions on PHP that will illuminate its sleek, modern feel, and we'll have something to talk about.

KV

2.4 What's in a Cache Miss?

> *People who are really serious about software should make their own hardware.*
>
> Alan Kay

Many people who develop software would like to pretend that hardware doesn't exist at all and that all their creations run on a perfectly imagined concept of a computer, but of course this is not actually the case. Software runs on hardware, and hardware is made up of bits that have to obey real-world constraints, such as the speed of light and entropy and other messy things that can interfere with our concepts of software.

While it might not be necessary to understand hardware at the level of the electrical engineers who build our chips, it is necessary at some level to understand how hardware affects the performance of software. No single change in computer architecture has had as profound an effect on overall system performance as that of introducing multiple levels of caching into CPUs. Most software performance is still predicated on a very old model of how CPUs execute software, but that model is long gone, except in perhaps the cheapest of low-end processors. The following letter and response tries to catch us up with the state of the art in which many of our performance problems lie hidden.

Dear KV,

I've been reading some pull requests from a developer who has recently been working in code that I also have to look at from time to time. The code he has been submitting is full of strange changes that he claims are optimizations. Instead of simply returning a value such as 1, 0, or -1 for error conditions, he allocates a variable and then increments or decrements it, and then jumps to the return statement. I haven't bothered to check whether or not this would save instructions, because I know from benchmarking the code that those instructions are not where the majority of the function spends its time. He has argued that any instruction we don't execute saves us time, and my point is that his code is confusing and hard to read. If he could show a 5 or 10 percent increase in speed, it might be worth considering, but he has not been able to show that in any type of test. I've blocked several of his commits, but I would prefer to have a usable argument against this type of optimization.

Pull the Other One

Dear Pull,

Saving instructions, how very 1990s of him. It's always nice when people pay attention to details, but sometimes they simply don't pay attention to the right ones. While KV would never encourage developers to waste instructions, given the state of modern software, it does seem like someone already has. KV would, as you did, come out on the side of legibility over the saving of a few instructions.

It seems that no matter what advances are made in languages and compilers, there are always programmers who think they're smarter than their tools, and sometimes they're right about that, but mostly they're not. Reading the output of the assembler and counting the instructions may be satisfying for some, but there had better be a lot more proof than that to justify obfuscating code. I can only imagine a module full of code that looks like this:

```
if (some condition) retval++; goto out: else retval-; goto out: ... out:
return(retval)
```

and, honestly, I don't really want to. Modern compilers, or even not so modern ones, play all the tricks that programmers used to have to play by hand: inlining, loop unrolling, and many others, and yet there are still some programmers who insist on fighting their own tools.

When the choice is between code clarity and minor optimizations, clarity must nearly always win. A lack of clarity is the source of bugs, and it's no good having code that's fast and wrong. First the code must be right, and then the code must perform; that is the priority that any sane programmer must obey. Insane programmers, well, they're best to be avoided. Eventually they wind up moving to a Central American nation, mixing their own drugs in bathtubs, and claiming they can unlock iPhones.

The other significant problem with the suggested code is that it violates a common coding idiom. All languages, including computer languages, have idioms, as pointed out at length in *The Practice of Programming* by Brian W. Kernighan and Rob Pike (Addison-Wesley Professional, 1999), which I recommended to readers more than a decade ago. Let's not think about the fact that that book is still relevant, and that I've been repeating myself every decade. No matter what you think of a computer language, you ought to respect its idioms for the same reason that one has to know idioms in a human language; they facilitate communication, which is the true purpose of all languages, programming or otherwise. A language idiom grows organically from the use of a language. Most C programmers, though not all, of course, will write an infinite loop in this way:

```
for (;;)
```

or as

```
while (1)
```

with an appropriate `break` statement somewhere inside to handle exiting the loop when there is an error. In fact, checking the *Practice of Programming* book, I find that this is mentioned early on (in section 1.3). For the return case, you mention it is common to return using a value such as 1, 0, or -1 unless the return encodes more than true, false, or error. Allocating a stack variable and incrementing or decrementing and adding a goto is not an idiom I've ever seen in code, anywhere, and now that you're on the case, I hope that I never have to.

Moving from this concrete bit of code to the abstract question of when it makes sense to allow some forms of code trickery into the mix really depends on several factors, but mostly on how much speedup can be derived from twisting the code a bit to match the underlying machine a bit more closely. After all, most of the hand optimizations you see in low-level code, in particular C and its bloated cousin C++, exist because the compiler cannot recognize a good way to map what the programmer wants to do onto the way the underlying machine actually works. Leaving aside the fact that most software engineers really don't know how a computer works, and leaving aside that what most of them were taught if they were taught about computers, hails from the 1970s and 1980s before superscalar processors and deep pipelines were a standard feature of CPUs, it is still possible to find ways to speed up by playing tricks on the compiler.

The tricks themselves aren't that important to this conversation; what's important is knowing how to measure their effects on the software. This is a difficult and complicated task. It turns out that simply counting instructions as your co-worker has done doesn't tell you very much about the runtime of the underlying code. In a modern CPU the most precious resource is no longer instructions, except in a very small number of compute-bound workloads. Modern systems don't choke on instructions; they drown in data. The cache effects of processing data far outweigh the overhead of an extra instruction or two, or ten. A single cache miss is a 32-nanosecond penalty, or about 100 cycles on a 3-GHz processor. A simple MOV instruction, which puts a single, constant number into a CPU's register, takes one-quarter of a cycle, according to Agner Fog at the Technical University of Denmark.

`http://www.agner.org/optimize/instruction_tables.pdf`

That someone has gone so far as to document this for quite a large number of processors is staggering, and those interested in the performance of their optimizations might well lose themselves in that site generally.

`http://www.agner.org`

The point of the matter is that a single cache miss is more expensive than many instructions, so optimizing away a few instructions isn't really going to win your software any speed tests. To win speed tests you have to measure the system, see where the bottlenecks are, and clear them if you can. That, though, is a subject for another time.

KV

2.5 Code Spelunking

> *Fools ignore complexity; pragmatists suffer it; experts avoid it; geniuses remove it.*
>
> Alan Perlis

A career in software means a career spent reading and trying to understand other koder's kode. The term *code spelunking*, which I coined in this article, tries to give a feel for what this process is like, to wit, crawling around a very large, dangerous space in the dark with primitive tools and a very small lamp. Many years have passed since I wrote this piece, and yet the tools used to spelunk code have not really kept pace with the amount of code any one person might need to dig through; in fact, the amount of code one would need to read, and the number of computer linguistical boundaries one has to cross has multiplied significantly.

A tool that is now sorely needed but that seems to elude us is one that provides useful visualization of a large code base. Most of the current crop of code spelunking tools, be they standalone or part of an IDE, allow a programmer to skip through the code in order to dive down, or up, a call chain. Systems like Doxygen have a rudimentary ability to produce a visual call graph, but these are static, hard to navigate, and quickly fail at scale. If there is one tool that most koders would use every day, it is indeed this sort of tool, because what most of us do, day to day, is dive into a morass of multiplying functions and try not to drown. The following piece is a tour of tools that might keep you afloat.

Try to remember your first day at your first software job. Do you recall what you were asked to do, after the human resources people were done with you? Were you asked to write a piece of fresh code? Probably not. It is far more likely that you were asked to fix a bug, or several, and to try to understand a large, poorly documented collection of source code.

Of course, this doesn't just happen to new graduates; it happens to all of us whenever we start a new job or look at a new piece of code. With experience we all develop a set of techniques for working with large, unfamiliar source bases. This is what I call code spelunking.

Code spelunking is very different from other engineering practices because it is done long after the initial design and implementation of a system. It is a set of forensic techniques used after the crime has been committed.

There are several questions that code spelunkers need to ask, and tools are available to help them answer these questions. I will look at some of these tools, addressing their shortcomings and pointing out possible improvements.

Source bases are already large, and getting larger all the time. At the time of this writing, the Linux kernel, with support for 17 different processor architectures, is made up of 642 directories, 12,417 files, and more than 5 million lines of code. A complex network server such as Apache is made up of 28 directories and 471 files—encompassing over 158,000 lines of code—whereas an application such as the nvi editor contains 29 directories, 265 files, and over 77,000 lines of code. I believe that these examples are fairly honest representations of what people are confronted with when they start a job.

Of course, even larger systems exist for scientific, military, and financial applications, but those discussed here are more familiar and should help convey an instinctive feeling for the complexity involved in systems that we all come in contact with every day.

Unfortunately, the tools we have to work with often lag behind the code we are trying to explore. There are several reasons for this, but the primary one is that very few companies have been able to build a thriving business on tools. It is much easier to make money selling software that appeals to a broad audience, which software tools do not.

Static vs. dynamic. We can use two scales to compare code spelunking tools and techniques. The first one ranges from static analysis at one end to dynamic analysis at the other. In static analysis you're not observing a running program but are examining only the source code. A clear example is using the tools find, `grep`, and wc to give you a better idea of the size of the source code. A code review is an example of a static technique. You print the code, sit down at a table with it, and begin to read.

On the dynamic end of the scale are tools such as debuggers. Here you run the code on real data, using one program (the debugger) to examine another as it executes. Another dynamic tool is a software oscilloscope, which displays a multithreaded program as a series of horizontal time lines—like those on a real oscilloscope—to find deadlocks, priority inversions, and other bugs common to multithreaded programs. Oscilloscopes are used mostly in embedded systems.

Brute force vs. subtlety. The second type of scale measures the level of finesse that is applied to spelunking. At one extreme is a brute-force approach, which often consumes a lot of CPU time or which might generate a large amount of data. An example of brute force is attempting to find a bug by using `grep` to locate a message that prints out near to where the error occurs.

At the other end of the finesse scale is subtlety. A subtle approach to finding a string within a program involves a tool that builds a database of all the interesting components of a code base (e.g., function name, structure definitions, and the like). You then use that tool to generate a fresh database each time you update your source from the code repository. You would also use it when you want to know something about the source, as it already has the information you need at its virtual fingertips.

Plotting your method. You can use both of these scales to create a two-dimensional graph in which the x-axis represents finesse and the y-axis runs from static to dynamic. You can then plot tools and techniques on this graph so that they can be compared.

None of the terms used implies value judgments on the techniques. Sometimes a static, brute-force approach is just what you want. If it would take just as long (or longer) to set up a tool to do some subtle analysis, a brute-force approach makes sense. You don't often get points for style when code spelunking; people just want results.

One thing to keep in mind when code spelunking is the Scout motto, "Be prepared." A little effort spent up front getting your tools in place will always save time in the long run. After a while, all engineers develop a stable of tools that they use to solve problems. I have a few tools that I always keep on hand when code spelunking, and I make sure to prepare the ground with them whenever I start a project. These tools include Cscope and global, which I'll discuss in more detail later.

A very common scenario in code spelunking is attempting to fix a bug in an unfamiliar piece of code. Debugging is a highly focused task: You have a program, it runs, but not correctly. You must find out why it does this, where it does this, and then repair it. What's wrong with the program is usually your only known quantity. Finding the needle buried in the haystack is your job, so the first question must be, "Where does the program make a mistake?"

You can approach the problem in several ways. The approach you choose depends on the situation. If the program is a single file, you could probably find the bug through inspection, but as demonstrated, any truly useful application is much larger than a single file.

Let's take a theoretical example. Jack takes a job with the Whizzo Company, which produces WhizzoWatcher, a media player application that can play and decode many types of entertainment content. On his first day at work (after he has signed up for health insurance, the stock plan, and the 401K) Jack's boss e-mails him two bug reports to deal with.

The two bugs that Jack has just been assigned are as follows:

Bug 1: When WhizzoWatcher opens a file of type X, it immediately crashes with no output except a core file. Bug 2: While watching a long movie on DVD (*The Lord of the Rings, The Two Towers*) the audio sync is lost after about two hours. This does not happen at any particular frame; it varies. WhizzoWatcher 1.0 is a typically organic piece of software. Originally conceived as a "prototype," it wowed the VPs and investors and was immediately rushed into production over the objections of the engineers who had written it. It has little or no design documentation, and what exists is generally inaccurate and out of date. The only real source of information on the program is the code itself. Because this was a prototype, several pieces of open source software were integrated into the larger whole and were expected to be replaced when the "real system" was funded. The total system now consists of about 500 files that spread over 15 directories, some of which were written in-house and some of which were integrated.

Bug 1. This is the easiest bug for Jack to work on; the program crashes when it starts. He can run it in the debugger, and the offending line will be found at the next crash. In terms of code spelunking, he doesn't have to look at much of the code at first, although becoming generally familiar with the code base will help him in the long run.

Unfortunately, Jack finds that although the reason for the immediate crash is obvious, what caused it is not. A common cause of crashes in C code is dereferencing a null pointer. At this point in the debugger, Jack doesn't have the previous states of the program, only the state at the moment it crashed, which is actually a very small amount of data. A common technique is to visually inspect the code while stepping up the stack trace to see if some caller stomped on the pointer.

A debugger that could step backward, even over a small number of instructions, would be a boon in this situation. On entry into a function the debugger would create enough storage for all of the function's local variables and the application's global variables so that they could be recorded, statement by statement, throughout the function. When the debugger stopped (or the program crashed), it would be possible to step backward up to the start of the function to find out which statement caused the error. For now, Jack is left to read the code and hope to stumble across the real cause of the error.

Bug 2. Attacking Bug 2, where the code doesn't crash but only produces incorrect results, is more difficult because there isn't a trivial way to get a debugger to show you when to stop the program and inspect it. If Jack's debugger supports conditional breakpoints and watchpoints, then these are his next line of defense. A watchpoint or a conditional breakpoint tells the debugger to stop when something untoward happens to a variable, and allows Jack to inspect the code at a point closest to where the problem occurs.

Once Jack has found the problem, it's time to fix it. The key is to fix the bug without breaking anything else in the system. A thorough round of testing is one way to address this problem, but he'll feel more comfortable making a fix if he can find out more about what it affects in the system. Jack's next question ought to be, "Which routines call the routine I want to fix?"

Attempting to answer this question with the debugger will not work, because Jack cannot tell from the debugger all the sources that will call the offending routine. This is where a subtle, static approach bears fruit. The tool Cscope builds a database out of a chunk of code and allows him to do the following:

Find a C symbol

Find a global definition

Find functions called by a function

Find functions calling a function

Find a text string

Change a text string

Find an `egrep` pattern

Find a file

Find all files that `#include` this file

You will note that item 4, "Find functions calling a function," answers his question. If the routine he's about to fix modifies nonlocal variables or structures, he must next answer the question, "With which functions does my function share data?" This would never come up in a "properly designed program" (i.e., one written from scratch). Jack would never use a horde of global variables to control his program, because he knows what a nightmare it would be to maintain. Right?

Of course, this is code spelunking, so Jack is already past that point. Using a subtle tool such as Cscope again is tempting, but this turns out to be a case where brute force is his best hope. Generating a list of all the global references in a program (filename and line number) is certainly possible for a compiler, but none of them do this. Although this option would hardly be used in the creation of new code, it would make the process of debugging old code far easier. Jack will have to make do with a combination of find and `grep` to figure out where all these variables reside in the program.

Code spelunking isn't something you do only when debugging; it's how you perform a good code review, reverse-engineer a design document, and conduct a security audit.

During an age when many people use computers for financial and personal transactions, auditing code for security holes has (or should) become commonplace. In order to close these security holes, you have to know what the common attacks are, as well as which sections of the code are vulnerable to attack. Although the attacks are updated almost daily on the Bugtraq mailing list (`http://www.securityfocus.com`), what we're concerned with is finding them.

Consider the following example. Jill takes a job with a large bank that serves most of its customers electronically, over the Internet. On her first day her boss tells her that the bank is doing a security audit of its entire system, which is implemented on several different types of hardware and operating systems. One set of Unix servers handles incoming web traffic and then makes requests to a mainframe backend that actually manages the real money.

One possible security hole occurs when a program stores a user's private data in a way that makes it available to anyone with access to the machine. An example of this is storing a password as plain text in a file or registry key. If Jill had written the software or had access to the person who did, she could quickly find out where the program stores passwords simply by asking. Unfortunately, the person who wrote the login scripts was laid off when the bank moved its headquarters six months ago. Jill will have to find out how this is done without the author's help.

Unlike in the debugging case, few tools are available to tell Jill something as specific as, "Where does the program store data X?" Depending on the structure of the program, this could happen anywhere—and often does. Jill can tackle this problem with either a brute-force or a subtle approach. To do the former she would litter the code with debugging statements (i.e., `printfs` or language equivalent) to find out where the program stores the password. She probably has a few guesses to narrow the field from the entire program down to a few or a dozen locations, but this is still a good deal of work.

A subtler approach would be to run the program in the debugger, attempt to stop the program right after the user enters a password, and then follow the execution until the program performs the storage operation.

A typical way to attack a program is to give it bad input. To find these vulnerabilities, Jill must ask the question, "Where does the program read data from an untrustworthy source?" Most people would immediately think that any code dealing with network data would be vulnerable, and they would be right—in part. With the advent of networked file systems, the fact that the code is executing a `read()` statement does not imply that it is reading from a local (i.e., trustable) source.

Whereas Jack's debugging in my earlier example attempted to zero in on a problem, Jill's code audit is more of a "fan-out." She wants to see as much of the code as possible and understand how it interacts both with its own modules ("How is data passed around?") and with external entities ("Where do we read and write data?"). This can be an even more daunting task than finding a needle in a haystack and may be more like labeling all the individual straws. In this case, the most important thing for Jill to do is to find the most likely places that will cause problems—that is, the places that are executed most often.

There is a tool that, although not originally intended for this purpose, can help to focus Jill's efforts: a code profiler. For example, gprof was originally written to tell an engineer which routines within the program are using up all the CPU time and are, therefore, candidates for optimization. The program is run with a workload (let a user bang on it or have it service requests from the network), and then the output is analyzed. A profiler will tell Jill which routines are being called most often. These are obviously the routines to check first. There is no reason for Jill to pore over the large portion of the code that is called infrequently, while the routines called most often may have gaping holes.

Routines that pass bad arguments to system calls are another common security problem. This is most often exploited against network servers in an effort to gain control over a remote machine. Several commercial tools attempt to discover these kinds of problems. A quick and dirty approach is to use a system call tracer, such as ktrace or truss, to make a record of what system calls are executed and what their arguments are. This can give you a good idea of where possible errors lie.

Code spelunking is about asking questions. The challenge is to get your fingers around the right parts of the code and find the answers without having to look at every line (which is a near impossibility anyway). There is one tool I haven't yet mentioned in this article, and that's the one sitting inside your head. You can begin applying good engineering practices even though they weren't used to create the code you're spelunking.

Keeping a journal of notes on what you've found and how things work will help you to create, and keep in mind, a picture of how the software you're exploring works. Drawing pictures is also a great help; these should be stored electronically along with your notes. Good experiment design of the type you may have learned—and then promptly forgotten—in physics class is also helpful. Just beating on a piece of code until it works is not an efficient way of figuring it out. Setting up an experiment to see why the code behaves a certain way may take a lot of thought, but usually very little code.

Finally, one of my favorite tools is the "stupid programmer trick." You get a colleague to look at the code, and then you attempt to explain to him or her what the code does. Nine times out of ten your colleague won't even need to say anything. Fairly quickly you realize what's going on, smack yourself on the forehead, say thank you, and go back to work. Through the process of systematically explaining the code out loud, you have figured out the problem.

There is no one tool that will make understanding large code bases easier, but I hope that I've shown you a few ways to approach this task. I suspect you'll find more on your own.

Tool Resources

Global

http://www.gnu.org/software/global/

This is the tool I apply to every source base that I can. Global is really a pair of tools: gtags and htags. The first, gtags, builds a database of interesting connections based on a source tree in C, C++, Java, or YACC. Once the database is built, you can use your editor (both Emacs and vi are supported) to jump around within the source. Want to know where the function you're calling is defined? Just jump to it. The second tool is htags, which takes the source code and the database generated by gtags and creates an HTML-browsable version of the source code in a subdirectory. This means that, even if you don't use Emacs or vi, you can easily jump around the source code finding interesting connections. Building the database is relatively speedy, even for large code bases, and should be done each time you update your sources from your source-code control system.

Exuberant Ctags

http://ctags.sourceforge.net

Works on dozens of languages: Ant, Asm, Asp, Awk, Basic, BETA, C, C++, C#, Cobol, DosBatch, Eiffel, Erlang, Flex, Fortran, HTML, Java, JavaScript, Lisp, Lua, Make, MatLab, OCaml, Pascal, Perl, PHP, Python, REXX, Ruby, Scheme, Sh, SLang, SML, SQL, Tcl, Tex, Vera, Verilog, VHDL, Vim, YACC. Very useful in a multilanguage environment.

Cscope

http://cscope.sourceforge.net/

Cscope was originally written at AT&T Bell Labs in the 1970s. It answers many questions, such as: Where is this symbol? Where is something globally defined? Where are all the functions called by this function? Where are all the functions that call this function? Like Global, it first builds a database from your source code. You can then use a command-line tool, Emacs, or one of a few GUIs written to work with the system to get your questions answered.

gprof

`https://ftp.gnu.org/old-gnu/Manuals/gprof-2.9.1/html_mono/gprof.html`

This is the standard profiling tool on most open source Unix systems. Its output is a list of routines ordered by how often they were called and how much of the program's CPU time was spent executing them. This can be useful in figuring out where to start looking for security holes in a program.

ktrace

This is a standard tool on open source operating systems. The name stands for "kernel trace." It will give you a listing of all the system calls made by a program. The output includes the arguments and return values given to and received from the calls. You use ktrace by running a program under it and then dumping the output with kdump. It is available on all open source Unix operating systems.

DTrace

First developed on Solaris, but now available on FreeBSD, Linux, and Windows. The best reference on this tool is still the book by Brendan Gregg and Jim Mauro.[1]

Valgrind

`https://valgrind.org`

Useful for finding all kinds of memory leaks in C and C++ programs.

---

[1]. Brendan Gregg/Jim Mauro: *DTrace: Dynamic Tracing in Oracle Solaris, Mac OS X, and FreeBSD*, 2011.

2.6 Input Validation

> *If builders built buildings the way programmers wrote programs, then the first woodpecker that came along would destroy civilisation.*
>
> Gerald Weinberg

Of all the stupid things, and there are many in software and security, perhaps the most stupid is the continuing inability of koders to remember to validate their input. From cross-site scripting to SQL injection attacks to so many other things, not properly validating input leads to a whole host of problems on your hosts. Perhaps my favorite reference on this is from Randall Munroe's xkcd, in which he writes about a child named, in part, "Drop Tables"; see `https://xkcd.com/327/`. There are now libraries and guidelines galore on this topic, for every computer language you can imagine; all that is required is the will to use them.

Dear KV,

I work for a company the builds all kinds of different web applications. We do everything from blogs and news sites to mail and financial systems. It depends on what the customer wants.

Right now our biggest problem at work is the number of bugs we have that relate to input validation. These bugs are totally maddening because each time one of them is fixed some other problem pops up in the same code, and the checking code is getting very close to spaghetti. Is there any way out of this tangle without some mythical technology, like natural language understanding?

Input Invalid

Dear II,

You've come across one of the biggest programming problems since the day we stupidly let non-engineers, i.e., users, touch our nice toys. Of course, computers aren't really very useful if they don't do something for actual people, but it is a pain. Systems would be so much cleaner without people. Alas, user input is a fact of life, and one that we all have to work with every day. User input is also one of the biggest sources of security holes in software as any reader of the Bugtraq mailing list can tell you.

The first rule of handling user input is, *"Trust no one!"* in particular your users. Although I'm sure 90% of them are perfectly nice people who go to their religious shrine of choice at the appointed time every week, or whatever it is perfectly nice people do, I don't actually know any perfectly nice people, but I have heard about them; nevertheless, there are the usual minority of thieves, jerks, and just plain idiots who will look at your nice web form as a place to steal money, play tricks, and generally cause havoc. The rest of the people, the perfectly nice ones, whom I've never met, won't actually attack your system, they'll just use it in a way they think is logical, and if their logic and your logic do not match, kaboom. Kode Vicious hates kabooms; they mean late nights and complaints from my doctor about alcohol and caffeine intake. I can't help it if he's stingy with the prescription meds, but let's not get into that now.

The second rule is, *"Don't trust yourself!"* This is another way of saying that you should check your results to make sure you're not missing anything. Just because you sent something to the user does not mean that they didn't do something a bit odd to it before it came back to you. A quick example is a web form. If you depend on the data you sent in a web form to the user, you had better check the whole form, and not just the parts you expected the user to change with their browser. It's a simple trick to exploit an error in form submission code by sending a slightly changed form with proper user input.

It sounds, from your description, as if the system you're using was written using what is called a black list. A black list is a set of rules that says which things are bad. During the Cold War the United States maintained black lists to prevent people it didn't like from getting jobs. Your name is on the list, sorry, no job. In the same way, software uses black lists to say which types of operations, in this case user input, are bad. The problem with black lists is that they are hard to maintain. They start off simple enough, saying things like, "Do not accept input with URLs in them." But they quickly get out of hand, with lists of the names for "JavaScript," of which there are many, and different types of tags to check for, and, and, and... I hope you get the idea. It is better to use white lists where this is possible.

A white list, unsurprisingly, is the opposite of a black list. White lists only contain the things that are allowed, and are often very short. An example is "accept only ASCII alphabetic characters." White lists can be overly restrictive, but they have a distinct advantage over black lists in that the only time you have to change a white list is to make it more permissive. A black list is, by default, mostly permissive, with the few exceptions that are the entries in the list.

My recommendation is to switch to using white lists, and to be very restrictive in what can be given to you by the user. Initially this seems a bit draconian, but it's probably the best way to protect your code, both from users and from turning into spaghetti.

KV

2.7 Dickering with Docs

> *Documentation is like sex; when it's good, it's very, very good, and when it's bad, it's better than nothing.*
>
> Dick Brandon

It is a long-held belief that koders hate documenting their systems: "If it was hard to write, it should be hard to understand." The fact is quite different because anyone who has had to look at a mass of undocumented code knows the value of good documentation.

One of the reasons that so many people struggle with documenting their code and documenting their designs is that they don't appreciate the fact that every good piece of writing, fiction or non, function block or hardware feature, requires a narrative. Without a narrative you just wind up splatting words on a page, rather than communicating anything useful to the reader. What's ironic is that in software we have functions, which generally take something as input and transform it to some form of output, and if that's not subject to the concept of narrative, I'm not sure what is. A good narrative, other than in surrealist fiction, takes the reader from a place of ignorance to a place of enlightenment. If you are documenting a single function, this can be as simple as answering the following four questions:

- What is the input?
- What is the expected transformation of the input?
- What form should the output arrive in?
- What are the valid and invalid (error) return values?

For concepts at a higher level than a single function we still need to take the reader from a place of ignorance to one of enlightenment. A major failing of documentation is to assume that the reader has knowledge that they do not, which does not mean that we belabor them with details from the dawn of time, but it is very important to remember that even if the reader is a future version of yourself, they need to have some context that you currently have in your head and that you need to write down so that what you are trying to explain makes sense. In short, you must consider your assumptions when writing documentation. For software an unexpressed, or uncoded, assumption is a bug, one that might even be caught by the compiler, but in documentation the only way that the reader will notice the bug of your unwritten assumption is when they trip over it while trying to use the system, or to debug or extend it in some way. A good way to get your assumptions out of your head is to write a glossary for your document, because this will force you to define your terms and to expand your acronyms, both of which are good sources of hidden assumptions. If every koder who had to document, or interact with a tech writer, thought just a bit more about narrative and what their assumptions are, the state of software overall would be vastly improved.

Good documentation, like good code, has to be maintained after it is written. Changes to the code usually require changes to the docs, unless the code change was to fix a bug and the bug fix brought the code closer to the original intent of the docs. Docs, like tests, must be kept in good sync with the code, or they quickly become useless or worse, point the reader off in completely the wrong direction.

Dear KV,

What do you think of systems such as Doxygen that generate documentation from code? Can they replace handwritten documentation in a project?

Dickering with Docs

Dear Dickering,

I'm not quite sure what you mean by "handwritten" documentation. Unless you have some sort of fancy mental interface to your computer that I have not yet heard of, any documentation, whether in code or elsewhere, is handwritten or at least typed by hand. If you're using anything else to type on your keyboard, please, keep it to yourself.

I believe what you're actually asking is if systems that can parse code and extract documentation are helpful, to which my answer is, "Yes, but..."

Any sort of documentation extraction system has to have something to work with to start. If you believe that extracting all of the function calls and parameters from a piece of code is sufficient to be called documentation, then you are dead wrong, but, unfortunately, you would not be alone in your beliefs. Alas, having beliefs in common with others does not make those beliefs right. What you will get from Doxygen on the typical, uncommented code base is not even worth the term "API guide." It is actually the equivalent of running a fancy `grep` over the code and piping that to a text formatting system such as TeX or troff.

For code to be considered documented there must be some set of expository words associated with it. Function and variable names, descriptive as they might be, rarely explain the important concepts hiding in the code, such as "What does this damnable thing actually do?" Many programmers claim their code is self-documenting, but, in point of fact, self-documented code is so rare that I am more hopeful of seeing a unicorn giving a ride to a manticore on the way to a bar. In fact, if I ever do see this, I will be both less surprised and quite happy, because it will mean that I'm in an excellent frame of mind. The claim of self-documenting code is simply a cover-up for laziness. At this point, most programmers have nice keyboards and should be able to type at 40–60 words per minute, and some of those words can easily be spared for actual documentation. It's not like we're typing on ancient, line printing terminals.

The advantage you get from a system like Doxygen is that it provides a consistent framework in which to write the documentation. Setting off the expository text from the code is simple and easy, and this helps encourage people to comment their code. The next step is to convince people to make sure that their code matches the comments. Stale comments are sometimes worse than none at all because they can misdirect you when looking for a bug in the code. "But it says it does X!" is not what you want to hear yourself screaming after hours of staring at a piece of code and its concomitant comment.

Even with a semi-automatic documentation extraction system, you still need to write documentation, because an API guide is not a manual, even for the lowest level of software. How the API's documentation comes together to form a total system and how it should and should not be used are two important features in good documentation and are the things that are lacking in the poorer kind. Once upon a time I worked for a company whose product was relatively low level and technical. We had automatic documentation extraction, which is a wonderful first step, but we also had an excellent documentation team. That team took the raw material extracted from the code and then extracted, sometimes gently and sometimes not so gently, the requisite information from the company's developers so that they could not only edit the API guide, but then write the relevant higher-level documentation that made the product actually usable for those who had not written it.

Yes, automatic documentation extraction is a benefit, but it is not the entire solution to the problem. Good documentation requires tools and processes that are followed rigorously in order to produce something of value both to those who produced it and to those who have to consume it.

KV

2.8 What's in the Foo Field?

Incorrect documentation is often worse than no documentation.

Bertrand Meyer

While we're dickering with docs we come to a very common problem in technical documentation, which is when the authors of the docs blindly take the comments from the code and turn them into the manual, or other documentation, leading to the situation where the documentation may well tell us that setting the value of a variable or field, A, to 1 will result in B being cleared, but without telling us why.

Good technical documentation doesn't just need a narrative, as discussed previously in Section 2.7, but must also help the reader to understand the why and the how of the system they are attempting to use, integrate, or build upon. The following letter and response attempt to give documentation authors a small set of hints as to what can make the lives of those who read their documentation a bit less frustrating.

Dear KV,

When will someone write documentation that tells you what the bits mean rather than what they set? I've been working to integrate a library into our system, and every time I try to figure out what it wants from my code, all it tells me is what a part of it is: "This is the foo field." The problem is that it doesn't tell me what happens when I set foo. It's as if I'm supposed to know that already.

Confoosed

Dear Confoosed,

Nowhere is this problem more prevalent than in hardware documentation. I am sure Dante listed a special ring of hell for people who document this way, telling you what something is while never explaining the why or how.

The problem with that approach is assumed knowledge. Most engineers, of both the hardware and/or software persuasion, seem to assume that the people they're writing documentation for, if they write documentation at all, already have the full context of the widget they're working on in their heads when they start to read the docs. The documentation in this case is a reference, but not a guide. If you already know what you need to know, then you're using a reference; if you don't know what you need to know, then you need a guide. Companies that care about their documentation will, at this point, hire a decent technical writer.

The job of a technical writer is to tease out of the engineer not only the what of a device or piece of software, but also the why and the how. It is a delicate job, because given the incredible malleability of software, one could go on for thousands of pages about the what, not to mention the why and how. The biggest problem is that the *what* is the easiest question to answer, because it is in the code when dealing with software, or the VHDL (VHSIC Hardware Description Language) when dealing with hardware. The what can be extracted without talking to another person, and who really wants to spend the day pulling engineers' teeth to get coherent explanations about how to use their systems? Since it is easiest to get at the what, most documentation concentrates on this part, often to the exclusion of the other two. Most tutorial documentation is short, and then at some point the rest of it is left as "an exercise to the reader." And exercise it is. Have you ever tried to lift a reference manual?

Although many engineers and engineering managers now give lip service to the need for "good documentation," they continue to churn out the same garbage that technical people have joked about since IBM intentionally left pages blank. A good writer knows that his or her job is to form in the mind of the reader a sense and an image of what the writer is trying to communicate. Alas, programmers and engineers have rarely been known as good writers; in fact, they are most often known as atrocious writers. It turns out that writers often want to relate, in some way, to people. That is, however, not something often said about technical folk, and in fact, it's often quite the opposite. Most of us want to go off into a corner and "do cool stuff" and be left alone. Unfortunately, none of us works in a vacuum, and so we must at least learn to communicate effectively with others of our ilk, if only for the sake of our own project deadlines.

Every software and hardware developer should be able to answer the following questions about systems they are developing:

1. Why did you add this? (field, feature, API)
2. How is this field, feature, or API used? Give an example.
3. Which other fields, features, and APIs are affected by using the one you are describing?

And if the answer to 1 is, "Management told me to," then it's time to fire management, or find a new job.

KV

2.9 Testy Tester

> *Testing is the process of comparing the invisible to the ambiguous, so as to avoid the unthinkable happening to the anonymous.*
>
> James Bach

It's a close competition between documentation and testing for which is the least liked by many koders, and changes in development processes over the years, such as test-oriented development, have done little if anything to change this attitude. Under time pressure, and pretty much everyone is under time pressure, most programmers prefer to write code rather than tests. On an intellectual level many know this is the wrong approach, but on a gut level, given the choice, it's always going to be the lines of code over the number of tests.

One of the larger challenges for koders who do not spend all of their time writing tests is knowing where to start. Here I lay out some simple guidelines for good test code, which ought to at least get most people started down the right path.

Perhaps one of the most important things about testing has to do with mindset or approach. A good first step is to get into the habit of thinking of testing as an application of the scientific method (see Section 2.1). The next step, and this is very difficult for most people, is to lose your assumptions about your own code. The second step is considered so difficult that most people think it's impossible and instead prefer to swap code between people who develop the code and those who develop the tests. The challenge with having two different people develop tests is that many companies think they can hire less experienced, aka cheaper, developers to write the tests, which is a grave error. The fact is that test and code development are equally challenging, and the smarter way to handle the problem of assumptions is to have koders of equal quality work part time on code and part time on tests, and to swap their work. Alice writes module A tested by Bob, and Bob writes module B tested by Alice. The role of the QA team should be to provide overall test infrastructure as well as filling in holes where critical tests are missed, and, even if Alice and Bob are excellent at their work, they're still going to miss things.

Dear KV,

I just joined a company building a large web services platform, and I'm working with their QA group. My current job is to write unit tests for the system and hook them into our nightly regression suite. A lot of the kode jockeys on the team complain that my tests are pointless and that I'm just wasting my time. These koders don't actually write tests themselves, so how do they know? I'm getting a bit tired of being hammered on all the time by these guys. Do you write tests or do you just code? How do you know that the test you write is a good one?

Testy Tester

Dear Testy,

First of all, all good koders write tests. No koder worth their paycheck would just crank out code all day without bothering to see if it worked! So, in answer to your first question, yes, I write tests. Actually, I secretly enjoy writing tests for other people's code as well as my own. Writing tests for other people's code is one of the more interesting ways to learn about a system and how a coder thinks. Creating tests for my own code is simply a way of making sure I don't look like an idiot. I hate looking like an idiot.

The real meat of your question could take far more time to answer than I have here, so I'll give a short lesson in what I personally consider to be good test creation. Perhaps the easiest set of tests to create, and the ones I think most koders are familiar with, are the tests you write after someone has discovered you've made a slight, well, let's call it a mistake, to be nice. You simply go through the bug database and create a test for every bug that exists, hook them into an automated harness of some sort, and away you go. This is a deceptively easy way to make yourself look good in front of your boss. You have a large body of work to point to, and you can show that the product doesn't break in the way that it broke before. It's my belief, and I know that others disagree, that these kinds of tests ought to be written by the koder fixing the bug, instead of by someone outside of the koding team. You broke it? You fix it! You test it! You make sure it damn well doesn't break again, period. Unfortunately, by my giving that work to the koders, I've taken the work away from the QA team, sorry.

I'm not exactly sure, from your letter, what the problem is that your co-workers see in your tests, but I can tell you the things that irk me when I have worked with test teams in the past. The first of these is silly test syndrome, STS for short. STS is usually caused by managers who still believe in measuring work by lines of code, as opposed to the quality of work. These managers demand that everything be tested, without ever looking at what should be tested, or at weighing risks. So, what happens is that the QA team goes off and beavers away writing a test for every possible knob in the system, starting from some arbitrary point A, and working until the product ships. Eventually the tests become so cumbersome that they take all night to run, and they rarely turn up any nasty issues. The reason this is ineffective is that the QA team is not allowed to use their brains, an important asset, to write tests that will find the bugs that will be the worst for the users of the system. Avoiding STS requires a few things. It requires a brain, which I believe you have, because you managed to write to me. The second thing it requires is a working relationship with the people designing and implementing the system. You have to be able to ask questions so that you can direct your efforts at the areas that are the riskiest. It makes little sense to test a library interface that does something simple and well understood when there are 10,000 lines of experimental, or just plain weird, code that is the company's secret sauce that also needs to be tested. The last thing that is required to avoid STS is a manager, or management chain, that understands what it means to do good tests. If you're working for one of these LOC folks, you're at a disadvantage, and you'll have to work around them to get good tests written. After a few months, though, writing good tests will pay off because you'll be the person in the group finding the most, and nastiest, issues.

The other thing that irks me, and that might be the source of the complaints, are random tests, scattered around, that are not coherently organized. I'm a very strong believer in good test harnesses. Make it easy for coders to write tests and they'll write them for you; they'll even run them before they ship the code. If your tests are sitting in your home directory and require a lot of setup, you can forget getting any help working with them. Tests should not be one-offs; they should be part of a system, just like the software that you're testing.

So, if you have either STS or your tests are just bits and pieces, you know why your co-workers are complaining. In my experience the skills required to write good tests are somewhat different, but just as worthy, as those required to write good code. For the most part it takes brains, curiosity, and a penchant for breaking things to write good tests.

KV

2.10 How to Test

> *Optimism is an occupational hazard of programming: feedback is the treatment.*
>
> Kent Beck

After all of KV's railing on testing over the years someone finally wrote in to ask, "What is a good test?" which seemed a fair point. The example in the following response is of testing a network firewall, but the advice given here can be applied to any system; it just happened to be that I was working on a firewall at the time I received this letter, and figuring out what was wrong with it was at the top of my mind. Even more than koding, testing ought to be amenable to the scientific method, which we discussed in Section 2.1. Each test should be based on a hypothesis about the system and should show whether the hypothesis is true or false. The nice thing about automated test frameworks is that you now have a place to hang these hypotheses and the ability to run them in continuous integration.

Dear KV,

You write about the importance of testing, but I haven't seen anything in your columns on how to test. It's fine to tell everyone that testing is good, but some specifics would be helpful.

How Not Why

Dear How,

The weasel's way out of this response would be to say that there are too many ways to test software to give an answer in a column. After all, many books have been written about software testing. Most of those books are dreadful and, for the most part, also theoretical. Anyone who disagrees can send me an e-mail with their favorite book on software testing, and I'll consider publishing the list or trashing the recommendation. What I will do here is describe how I have set up various test labs for my specific type of testing, and maybe this will be of some use.

There are two requirements for any testing regimen: relevance and repeatability. Test-driven development is a fine idea, but writing tests for the sake of writing tests is the same as measuring a software engineer's productivity in KLOC. To write tests that matter, test developers have to be familiar enough with the software domain to come up with tests that will confirm that the software works and that also attempt to break the software. Much has been written about this topic, so I'm going to switch gears to talk about repeatability.

Tests are considered repeatable when the executions of two different tests on the same system do not interfere with each other. A concrete example from my own work is the population of various software caches such as routing and ARP tables that might speed up the second test in a series of tests of packet forwarding. To achieve repeatability, the system or person running the test must have complete control over the environment in which the test runs. If the system being tested is completely encapsulated by a single program with no side effects, then running the program repeatedly on the same inputs is a sufficient level of control. But most systems are not so simple.

Working from the concrete example of testing a firewall: To test any piece of networking equipment that passes packets from one network to another, you need at least three systems, a source, a sink, and the device under test (DUT, in test parlance). As I pointed out earlier, repeatability of tests requires a level of control over the systems being tested. In our network testing scenario, that means each system requires at least two interfaces and the DUT requires three. The source and sink need both a control interface and the interface on which packets will be either sent to or received from the DUT. "Why can't we just use the control interfaces to source and sink the packets?" I hear you cry. "Wiring all that stuff is complicated, and we have three computers on the same switch; we can just test this now." The way it works is that the control and test interfaces must be distinct on all the systems to prevent interference during the test. No matter what you are testing, you must make sure that you reduce the amount of outside interference unless that is what you are intending to test. If you want to know how a system reacts with interference, then set up the test to introduce the interference, but don't let interference show up out of nowhere. In our specific networking case, we want to retain control over all three nodes, no matter what happens when we blast packets across the firewall. Retaining control of a system under stress is nontrivial.

Another way to maintain control over the systems is to have access to a serial or video console. This requires even more specialized wiring than just a bunch more network ports, but it is well worth it. Often, bad things happen, and the only way to regain control over the systems is via a console login.

The ultimate fallback for control is the ability to remotely power-cycle the system being tested. Modern servers have an out-of-band management system, such as IPMI, that allows someone with a username and password to remotely power-cycle a machine, as well as do other low-level system management tasks including connecting to the console. Whenever someone wants me to test networked systems in the way I'm describing, I require them to have either out-of-band power management via a network-connected power controller or IPMI on the systems in question. There is nothing more frustrating during testing than having a system wedge itself and either having to walk down to the data center to reset it or, worse, having your remote hands have to do it for you. The amount of time I've wasted in testing because someone was too cheap to get IPMI on their servers or put in a proper power controller could have been far better spent killing the brain cells that had absorbed the same company's poorly written code. It seems that inattention to detail is pervasive, and when I see a poor testing setup, I should be prepared to see poor code as well.

At this point, we know that we have to retain control over the systems and we have several ways to do that via separate control interfaces, and ultimately, we have to have control over the system's power. The next place that most test labs fall down is in access to necessary files.

Once upon a time a workstation company figured out that they could sell lots of cheap workstations if they could concentrate file storage on a single, larger, and admittedly more expensive server. Thus was born the Network File System, the much-maligned, but still relevant, way of sharing files among a set of systems. If your tests can in any way destroy a system or if upgrading a system with new software removes old files, then you need to be using some form of networked file system. Of late I've seen people try to handle this problem with distributed version control systems such as git, where the test code and configurations are checked out onto the systems in the test group. That might work if everyone were diligent about checking in and pushing changes from the test system. But in my experience, people are never that diligent, and inevitably someone upgrades a system that had crucial test results or configuration changes on it. Using a networked file system will save whatever hair you have left on your head. (I should have learned this lesson sooner.) Make sure that the networked file system traffic goes across the control interfaces and not the test interfaces. That should go without saying, but in test lab construction, much of what I think could go without saying needs to be said.

At this point we have fulfilled the most basic requirements of a networking test system: We have control over all the systems, and we have a way to make sure that all the systems can see the same configuration data without undue risk of data loss. From here it's time to write the automation that controls these systems. For most testing scenarios, I tend to just reboot all the systems on every test run, which clears all caches. That's not the right answer for all testing, but it definitely reduces interference from previous runs.

KV

2.11 Leave the Test Modes In!

> *It's hard enough to find an error in your code when you're looking for it; it's even harder when you've assumed your code is error-free.*
>
> Steve McConnell

There are koders who love to add code, and there are koders who love to delete code. For some people there is nothing more satisfying than removing dead code, which is effectively the koder's version of weeding. In gardening, as in software, weeding is an important function of our work for it clears away unused code, reducing code size, complexity, and, from a security standpoint, attack surface. The problem comes when people cannot figure out if something is a weed or an important, and protective, part of the ecosystem.

The following letter and response cover the case of keeping in code that would purely be used for testing. Many systems have this sort of code protected under conditional compilation, but there are other systems that ship with the test code fully baked into the system. Leaving the test code baked in has some distinct advantages. Shipping the test code means that if there is a problem that needs to be debugged in production or at the point of use of the code, it's trivial to turn the test code on and run the tests. If the test code were not shipped, then a new binary would have to be sent along, and it's always possible that having the test code compiled in, vs. out, will result in a Heisenbug (see Section 2.16). But what if the code being left in is somehow dangerous if used in production, as in this next case? Well, let's see what I had to say about that.

Dear KV,

While reviewing some encryption code in our product, I came across an option that allowed for null encryption. This means the encryption could be turned on, but the data would never be encrypted or decrypted. It would always be stored "in the clear." I removed the option from our latest source tree because I figured we didn't want an unsuspecting user to turn on encryption but still have data stored in the clear. One of the other programmers on my team reviewed the potential change and blocked me from committing it, saying that the null code could be used for testing. I disagreed with her, since I think that the risk of accidentally using the code is more important than a simple test. Which of us is right?

NULL for Naught

Dear NULL,

I hope you're not surprised to hear me say that she who blocked your commit is right. I've written quite a bit about the importance of testing, and I believe that crypto systems are critical enough to require extra attention. In fact, there is an important role that a null encryption option can play in testing a crypto system.

Most systems that work with cryptography are not single programs, but are actually frameworks into which different cryptographic algorithms can be placed, either at build or at run time. Cryptographic algorithms are also well-known for requiring a great deal of processor resources, so much so that specialized chips and CPU instructions have been produced to increase the speed of cryptographic operations. If you have a crypto framework and it doesn't have a null operation, one that takes little or no time to complete, how do you measure the overhead introduced by the framework itself? I understand that establishing a baseline measurement is not common practice in performance analysis, an understanding I have come to while banging my fist on my desk and screaming obscenities. I often think that programmers shouldn't just be given offices instead of cubicles, but padded cells. Think of how much the company would save on medical bills if everyone had a cushioned wall to bang their heads against, instead of those cheap, pressboard desks that crack so easily.

Having a set of null crypto methods allows you and your team to test two parts of your system in near isolation. Make a change to the framework, and you can determine if that has speeded up or slowed down the framework overall. Add in a real set of cryptographic operations, and you will then be able to measure the effect the change has on the end user. You may be surprised to find that your change to the framework did not speed up the system overall, as it may be that the overhead induced by the framework is quite small. But you cannot find this out if you remove the null crypto algorithm.

More broadly, any framework needs to be tested as much as it can be in the absence of the operations that are embedded within it. Comparing the performance of network sockets on a dedicated loopback interface, which removes all of the vagaries of hardware, can help establish a baseline showing the overhead of the network protocol code itself. A null disk can show the overhead present in file-system code. Replacing database calls with simple functions to throw away data and return static answers to queries will show you how much overhead there is in your web and database framework.

Far too often we try to optimize systems without sufficiently breaking them down or separating out the parts. Complex systems give rise to complex measurements, and if you cannot reason about the constituent parts, you definitely cannot reason about the whole, and anyone who claims they can is bull$#!+ting you.

KV

2.12 Maintenance Mode

> *All programming is maintenance programming, because you are rarely writing original code.*
>
> Dave Thomas

I won't say that every koder comes out of school expecting to write new code in a green field, but that is how computer science courses continue to be taught. Students usually write code that is meant to be original, which makes it easier to grade, but which completely fails to train them in what their working lives will be like.

The day-to-day work of most people in software is one of working, for the most part, on the mountain of dung that is the current set of software stacks. We all add to this mountain but rarely are we able to walk away from it to create something wholly new. Given this fact, and the nasty smell wafting from that mountain, it's best if we learn to handle the Sisyphean tasks required to maintain the mountain so that parts don't roll down on top of us.

Of course, software maintenance would be easier if systems were designed to be well-maintained, which frequently they are not. Most software is designed and built *in the now*, that is, if it's designed at all. Building software that can be maintained requires that it has a clear design, that the implementation follows some basic patterns, and that the resulting code is readable by someone other than the original author.

Dear KV,

Isn't software maintenance a misnomer? I've never heard of anyone reviewing a piece of code every year, just to make sure it was still in good shape. It seems like software maintenance is really just a cover for bug fixing. When I think of maintenance, I think of taking my car in for an oil change, not fixing a piece of code. Are there any people who actually review code after it has been running in a production environment?

Still Under Warranty

Dear Still,

The short answer is that, yes, the term software maintenance is yet another computer industry bit of Newspeak, probably invented so that people wouldn't have to put "bug fixer" on their resumes. Since there is strength in ignorance, I should probably not answer your question any further, but I find it impossible now to follow the party line.

Although the term "software maintenance" has little meaning other than bug fixing, the question you ask has some deeper implications. The reason to do maintenance on a car, or any other machine, is that it has moving parts that wear out over time. Parts wear out because they are subjected to physical stresses, such as two gears, one of which drives the other in order to transfer energy from a motor to a wheel. Software isn't subjected to physical stresses, though there are pieces of code I've read, only just recently, that rightly deserve to be put under physical stress, or at least their authors do. Even though software itself does not wear out after it is executed repeatedly, there is a place for maintenance in the software industry.

The more modern term for what really is software maintenance is another bit of Newspeak: refactoring. It is unfortunate that every term we come up with to describe something simple and direct in our industry is almost instantly debased so that it loses all meaning. While I'm not generally against debasement, in this case it does make our lives a bit more difficult. Many people wrongly use the term refactoring to stand in for: "Oh, we really screwed up the entire system we were writing over the past six months, so we're going to have to write it again from scratch, but we'll keep the name of the program the same, as well as the names of many of the classes. We will, of course, have to change all the insides of the classes, their method signatures, and about 90 percent of the rest of the text that was the program, but from the outside it will look the same, except for all the features; those will change too." When you replace most of your API, the innards of the code, and the features, that is not refactoring; nor is it maintenance. That's a rewrite.

Refactoring actually means that you have some functions or classes that, with a small number of changes, can be used in another program. It is this more honest type of refactoring that begins to look like software maintenance. The reason that refactoring is more like maintenance is because you're looking at several moving parts that no longer mesh well with each other. In the previous program or system in which they were used, they meshed well; otherwise, that program would not have worked correctly. But they do not mesh correctly with the new design; therefore, you're required to add a metaphorical bit of grease, or break off a few teeth from the gears to make them work well in their new application.

When you're refactoring a piece of code is also the perfect time to think about the original design: if it made sense originally, if it makes sense now, and if you want to be stuck with this same design in the future. I am *not* saying that you should rework every single piece of code you see just to make it a bit cleaner, nicer, more generic, etc. You can't actually even imagine that I would give people license just to diddle with all the code in a system. That kind of navel-gazing really ought to result in whippings, but I hear that those are not allowed in most workplaces at this point.

While you're maintaining, oh, I mean refactoring, please remember that whatever you change must be tested. Just because you "changed the API only a little bit by adding one teensy-weensy little bit field" does not excuse you from testing your change. If you don't, I can guarantee that you'll be back doing the old kind of software maintenance (i.e., bug fixing).

KV

2.13 Merge Early

> *Merge early, merge often.*
>
> — Anon

The following letter was written before git won the current round of the War of the VCSs. In a world where git is the main version control system, all one thinks about, other than suicide, is merging, because that's the only mode of operation for git. Anyone who has been tempted to rebase in git knows that they'd be happier freebasing instead. The question answered herein isn't related directly to a particular VCS but deals more in the question of how often you want to share your work with your co-workers, as well as how often you're willing to incorporate their mistakes, I mean, changes, into your own.

2.13 Merge Early

Dear KV,

When doing merged development, how often should you merge? It's obvious that if I wait too long, then I spend days in merge hell, where nothing seems to work and where I wind up using the *revert* command more often than *commit*; but the whole point of branched development is to be able to protect the main branch of development from unstable changes. Is there a happy middle ground?

Merge Daemon

Dear Merge,

For years people have asked me about the "happy middle ground," not only in branched development but also in many areas. I don't know if you've noticed this, but the tenor of these questions is rarely happy and even more rarely results in anything that could be considered a middle ground.

There are two related themes in dealing with merge purgatory (if it were hell, there would be no chance of escape, but from purgatory you might be able to work your way out). The first theme is related to how the software you're working on is developed. If there are well-defined boundaries between the sections of the code, then you really should not have to enter merge purgatory at all, because if you're working on a component, then no one else should be making changes to it; the only differences will be yours, case closed and well done there. Of course, that's pretty unlikely in any project larger than a few people.

Projects often have the problem of everyone having their fingers in all of the code. That's not difficult to solve if the project is small: you can simply sit down with the other participants in the project and break up the work along some logical boundaries. In a large project you're likely to have several people looking into and modifying the same piece of code. Perhaps one is fixing bugs while another is developing new features, or there are two versions of the product or project that have different teams working from the same main line. Whatever the case, there are a few things you should do to avoid purgatory, such as having an automated build process so that everyone knows when something broken got merged or checked in. For more on this, see my response in Section 5.3.

The second and more relevant point to this discussion is the direction of merges. When you're working off in your own branch, it's kind of like working off in a corner of a building, usually a nice quiet place away from the hubbub of your co-workers, where you can concentrate on your work and get things done. Alas, while you're coding in this temporary Nirvana, the rest of the world continues on, with all its attendant suffering. It would be nice if you could remain in Nirvana, but, well, you can't.

You have two choices: You can decide to remain in Nirvana as long as possible and then at the very end stride boldly into the mouth of merge purgatory, or you can allow small bits of suffering to enter your world. The suffering, in this case, is a merge from the main branch, the one in which everyone is committing changes, and from which all of your suffering ultimately comes. When I'm working on branched development, and at this point that's about 95 percent of the time, I merge early, and I merge often, just like when I vote.

Now when I say "merge early," I do not mean "early in the morning." As a matter of fact, I pretty much never say "early in the morning" because I really don't believe such a time exists, and anyone who tries to tell me there is such a time I consider to be a figment of my imagination. In particular, I recommend against merging anyone else's code before you write or debug some of your own. Few things are more frustrating than to start your working day with someone else's broken code messing up your beautiful branch. Get some of the things you wanted to do done first and then integrate a bit of the main branch's suffering into your own. I prefer to do these kinds of merges after I feel good about having gotten some of my own work done. At least then I'm starting out on a high, and the drudgery of fixing my branch will bring me back to a neutral state instead of sending me down into a code depression. I would say that on a fast-moving project you should be merging from the main branch once a day, and no less than once a week. A lot can happen in a week, and you don't want to spend your weekend fixing your working branch.

When is it best to merge code back into the tree? That depends on what you're merging. If it's bug fixes, then they need to be merged as soon as they're tested because there are likely other people depending on those bug fixes. For features, they should be complete, which means tested and ready to use. Complete does not mean bug free. At some point you have to stop polishing that turd and just flush it into the system.

That covers the process part of when to merge, but there is also a tool component. Some source-code control systems simply do not have the tools to support branched development, while others encourage so much branching that code never comes back to the main line. Avoid both of these types of systems if you want to do branched development.

A good system for branched development must include a decent tool for comparing multiple versions of a file. In a better world it would be possible for the source-code control system to handle all merges automatically, but this is not yet possible, so the act of performing a merge is really one of handling the exceptions to the merging process.

If the tool that calculates the differences between files in different branches is a poor one, then a larger amount of work will fall to you to integrate the differences. The reason people call it "merge hell" isn't because of the merging, per se, but because of the amount of human intervention that is required. Do you want the code in version 1.2.2.1 that Bob wrote to be merged with your code in your current version? These are the types of decisions that make merging so unpleasant. On a positive note, this is a known problem, and many people are trying very hard to provide us with better tools, so it's one area in which things are actually improving.

To sum up, my advice is to integrate changes into your branch daily but not at the beginning of the day and to work with a modern source-code control system that has good support for merging changes and resolving differences.

KV

2.14 Multicore Manticore

> *Software gets slower, faster than hardware gets faster.*
>
> Niklaus Wirth

More than a decade ago the hardware industry told the software industry that we were going to run out of leaps in processing power and that to solve this problem the hardware industry would give the software people more processing cores rather than increasing the frequency of a single CPU. That was all they could do with the increasing numbers of transistors they could put on the chip. If frequencies had been able to increase from that point until now, we'd have 20 GHz processors, rather than still topping out just shy of 5 GHz, and those CPUs require water cooling.

The problem with the hardware folks telling the software folks to "go multicore" is that hardware people don't really understand distributed systems, networking, or locking problems, and neither do most software people. In order to take advantage of all those cores, problems have to be broken down in such a way that they don't step on each other's toes, and this is a hard and continuously unsolved problem. To this very day most software is still written in a single-threaded manner, even when it might be better off being written to take advantage of multiple threads. Elsewhere I talk about the problems of threaded programming, but here I just cover the multicore problem, at its core.

Dear KV,

In the past 10 years I've noticed that the number of CPU cores available to my software has been increasing, but that the frequency of those cores isn't much more than it was when I left school. Multicore software was a big topic when the trend first began, but it doesn't seem to be discussed as much now that systems often have six or more cores. Most programmers seem to ignore the number of cores and write their code as they did when systems had only a single CPU. Is that just my impression, or does this mean that maybe I picked the wrong startup this year?

Core Curious

Dear Core,

The chief contribution of multicore hardware to software design has been to turn every system into a truly concurrent system. A recently released digital watch has two cores in it, and people still "think digital watches are a pretty neat idea" (as in Douglas Adams' *The Hitchhiker's Guide to the Galaxy*). When the current crop of computer languages was written, the only truly concurrent systems were rare and expensive beasts that were used in government research labs and other similarly rarefied venues. Now, any clown can buy a concurrent system off the Internet, install it in a data center, and push some code to it. In fact, such clowns can get such systems in the cloud at the push of a button. Would that software for such systems were as easily gotten!

Leaving aside the fact that most applications are now distributed systems implemented on many-core communicating servers, what can we say about the concurrent nature of modern software and hardware? The short answer is, "It's all crap," but that's not helpful or constructive, and KV is all about being helpful and constructive.

From our formative computer science years, we all know that in a concurrent system two different pieces of code can be executing simultaneously, and on a modern server, that number can easily be 32 or 64 rather than just two. As concurrency increases, so does complexity. Software is written to be executed as a set of linear steps, but depending on how the software is written, it may be broken down into many small parts that might all be running at the same time. As long as the software doesn't share any state between the concurrent code, everything is fine, well, as fine as any other nonconcurrent software. The purpose of software is to process data and therefore to take things and change them or to mutate state. The number of significant software systems that do not wind up sharing state between concurrent parts is very, very small.

Software that is written specifically without concurrency is, of course, easier to manage and debug, but it also wastes most of the processing power of the system on which it runs, and so more and more software is being converted from nonconcurrent into concurrent or being written for concurrency from scratch. For any significant system, it is probably easier to rewrite the software in a newer, concurrency-aware language than to try to retrofit older software with traditional concurrency primitives.

Now, I'm sure you've read code that looks nonconcurrent, that is, it does not use threads in its process, and you might think that was fine, but, alas, nothing is ever really fine. Taking a collection of programs and having them share data through, for example, the file system or shared memory, a common early way of having some level of concurrency, does not protect the system from the evils of deadlock or other concurrency bugs. It's just as possible to deadlock software by passing a set of messages between two concurrent processes as it is to do the same sort of thing with Posix threads and mutexes. The problems all come down to the same things: idempotent updates of data and data structures, the avoidance of deadlocks, and the avoidance of starvation.

These topics are covered in books about operating systems, mostly because it was operating systems that first had these challenges. If, after this description, you're still curious, I recommend picking up one such book so that you at least understand the risks of concurrent systems and the land mines you are likely to step on as you build and debug, debug, and debug such systems.

KV

2.15 This Is Not a Product

> *If at first you don't succeed, call it version 1.0.*
>
> Pat Rice

How do you know when a system is done? Can you set a timer? Take its temperature and see if it's baked? Many companies like to field software that is of beta quality and see how the market responds to it, and whether or not the product has legs, as marketing people are wont to say, a saying that makes KV want to break their legs. Some companies, and here I'm thinking Google, would even field software for years and label it beta, as a way of gathering suckers, I mean users and data, in order to make the version after the beta. It is unfortunate when slick marketing, or monopolistic market share, allows companies to foist half-baked systems on users, even if the users would prefer a half-baked product to nothing at all. Alas, caveat emptor is as true in software as it is in any other business.

Dear KV,

After spending the last year buying and deploying a set of monitoring systems in my company's production network, we hit our first bug in the manufacturer's system software. After we reported the bug, the manufacturer asked us to upgrade the systems to the latest release. It turns out that the upgrade process requires us to reset the devices to their factory defaults, losing all our configuration information on each system and requiring a person to reenter all of the configuration data after the upgrade.

At first, we thought this might be something required only for this particular upgrade, which crosses a major revision, but it turns out that we will have to reenter our configuration data each and every time we upgrade the software on these systems. I guess we should have asked this question before we bought the systems, but it just never occurred to us that anyone would sell a box purporting to act as an appliance that could not be easily upgraded in the field. One of the other guys in my group suggested we just return the boxes and ask for our money back, but, depressingly enough, these are the best systems we've found for network monitoring.

Down on Upgrades

Dear Down,

It seems that what you thought was a product, that is, a set of components thoughtfully put together by people who care about their customers, was actually just a collection of parts that under normal circumstances worked well enough to get by. Unfortunately, the number of companies that think about what's going to happen after version 1.0 of their systems is quite small.

I've been fortunate, or perhaps I should say that the sales critters I've dealt with in the past 10 or so years have been fortunate not to come across too many systems that act in the way you describe. The idea that the manufacturer would require a user or systems administrator to reenter already saved data after an upgrade is stupid, ludicrous, and a bunch of other words that my editors just aren't going to allow in this publication.

Even in the worst-designed products (and I've used enough of those) there is usually some bit of Perl that takes the old configuration data and turns it into something that's mostly valid for the latest revision.

As a matter of fact, many years ago I worked on a networking switch project, and the first thing the systems team (which was responsible for getting a reasonable operating system and applications onto the box) did was come up with a way to field-upgrade the system. Anyone who has ever configured a switch or router knows that you don't just toss out the configuration on an upgrade.

The sad part is that doing this correctly just isn't that difficult. Most embedded systems now use stripped-down Unix systems such as Linux or the BSDs, all of which store their configurations in well-known files. Granted, this isn't the nicest way to store configuration data, because it tends to be a bit scattered, but it's not that hard to write a script that can handle differences between the versions and reconcile them. On FreeBSD there is etcupdate, and Linux has etc-update and dispatchconf. In a properly designed system the configuration would likely be stored in a simple database or an XML file, both of which are field-upgradable with fairly simple scripts.

These sorts of issues are what differentiate an appliance that can be deployed and maintained with little human interference from a system that has to be diddled constantly to keep it happy. It is a sad fact that many programmers and engineers do not think much of appliances and are more likely to think that their users should "man up" and spend their time making up for the original designer's lack of thought.

I still remember one of the first all-digital stereo systems I ever saw. It was designed around a Sun workstation and cost on the order of $15,000. It was obvious from the moment you looked at the controls that little or no thought had been put into what people really wanted out of a sound system. What most people want out of a stereo is good-quality sound with a minimum of button pushing to get what they want. What the system I saw presented was a lot of button pushing for about the same quality of sound that I could get out of an amplifier and a CD player. If the user interface was horrific, it was nothing compared with the system performance. The box crashed three times before I finally walked out of the store. The one thing I got from that experience was an understanding that systems and products are not the same thing.

A system is a collection of components put together to do a job. A product is a system that has been designed and built to make the work of carrying out the job smooth and natural to the user. I can certainly cobble together software to rip, store, and play my CDs on a computer; that is a system, whereas an iPhone is a product. When I upgrade my phone, I don't reenter my data. If I did, the product would have died at the first revision. More likely, Steve Jobs would have taken someone's head off if he had been told that the upgrade path required reentering data. Given how complex modern appliances are (and let's face it, your TV probably has an Ethernet jack), it's clear that people have thought about and solved this problem.

The real issue is with the people who design these devices. Somehow the fact that an appliance is going to be hanging out with a bunch of computers, for example, in a colocation, makes it acceptable for the implementers to make their box look more like an open source desktop, which will be diddled by an experienced IT person or the users themselves. It's a practice that really needs to stop, if only because I don't want everyone to wind up bald, like me. My hair didn't fall out, I pulled it out!

KV

2.16 Heisenbugs

> *Heisenberg may have slept here.*
>
> Anon

Werner Heisenberg wasn't a software engineer, but a physicist, and one of his major contributions was the uncertainty principle that states that the more we know about one part of one thing in a physical system, the less we can know about another, e.g., the position of a particle in space vs. its speed.

The Heisenbug is a nasty one, which is why it gets its own letter, but, unlike the problem in physics, software engineers can actually do things to avert these types of problems. There are a whole host of tools, both in hardware and in software, that can help address this type of problem. A little time learning to use an In Circuit Emulator, a JTAG interface, or even a logic analyzer can pay off when your software tools no longer have the fidelity required to catch one of these bugs. Most modern server CPUs have hardware watchpoints as well as various types of performance monitoring systems that can also be used to track these down.

Dear KV,

The company I work for rolled out a new monitoring system one weekend, and it didn't go as well as we would have liked. When we first brought up the monitoring system, several of our servers started to show very high CPU load. Initially, we could not figure out why. The monitoring processes on each server were very busy, so we turned off the monitoring system, and the servers got less busy. Eventually, we realized it was the number of polls being issued by the monitoring system that was causing the servers to use so much CPU time. We decreased the polling frequency to every 10 minutes, and this seemed to be the sweet spot for system performance. What I would like to know is how one should go about tuning such systems, as it seems still to be done via trial and error.

Polled Too Frequently

Dear Polled,

Trial and error? The problem here is usually a failure to appreciate just what you are asking a system to do when polling it for information. Modern systems contain thousands, sometimes tens of thousands, of values that can be measured and recorded. Blindly retrieving whatever it is that might be exposed by the system is bad enough, but asking for it with a high-frequency poll is much worse for several reasons.

The first reason is the one that you bring up in your letter: the amount of overhead introduced by simply asking for the data. Whenever you ask the system for its configuration state, whether that's a routing table or the state of various `sysctls` (system control variables), the system has to pause other work to provide a consistent picture of what's going on. KV knows that in recent years the idea of consistency has been downplayed in favor of performance in particular, by various database projects. In the systems world, however, we still think that consistency is a good thing™, and therefore the system will try either to snapshot the data you request or to pause other work while the data is read out. If you ask for a few thousand items and a random `sysctl -a` shows more than 9,000 elements on a server I am using, then that's going to take time, not forever but not nothing, either.

The second reason that polling for data frequently is a problem is that it actually hides the information you might be looking for in the noise generated by retrieving and communicating the values you asked for. Every time you ask the system for some stats, it has to do work to get those stats, and the system doesn't account for your request separately from any other work it has to do. If your monitoring system is banging away at the server asking for data every minute, then what you will see in your monitoring system is the load that the system itself is generating. Such Heisen-monitoring, where your monitoring system is overwhelmingly affecting the measurements, is completely pointless.

In a monitoring system, there is always the tension between too much and too little information. When you're debugging a problem, you always wish you had more data, but when your system is running normally, you want it to do the work for which it was deployed. Unless you get off on just pushing monitoring systems (and, yes, there is definitely a handle for those people somewhere on social media) you need to find the Goldilocks zone for your monitoring system. To find that zone, you must first know what you're asking for. Figure out which commands the monitoring system is going to execute on your servers, and then run them individually in a test environment and measure the resources they require. You care about runtime, which can be found to a coarse level with the `time(1)` command. Here is an example from the server just mentioned:

```
time sysctl -a > /dev/null sysctl -a > /dev/null
        0.02s user 0.24s system 98% cpu 0.256 total
```

Here, grabbing all of the system's various system-control variables takes about a quarter of a second of CPU time, most of which is system overhead, that is, time spent in the operating system getting the information you requested. The `time(1)` command can be used on any utility or program you choose.

Now that you have a rough guess as to the amount of CPU time that the request might take, you need to know how much data you're talking about. Using a program that counts characters, such as `wc(1)`, will give you an idea of how much data you're going to be gathering and moving off the system for each polling request.

```
sysctl -a | wc -c 378844
```

You would be grabbing more than a quarter of a megabyte of data here, which in today's world isn't much, but it still averages out to 6,314 bytes per second if you poll every minute; and, in reality, the instantaneous rate is much higher, causing a 3 Mbps blip on the network every time you request those values.

Of course, no one in his or her right mind would just blindly dump all the `sysctl` values from the kernel every minute; you would be much more nuanced in asking for data. KV has seen a lot of unsubtle things in his time, including monitoring systems that were set up to do just this sort of ridiculous level of monitoring. "We don't want to lose any events; we need a transparent system to find bugs!" I hear the DevOps folks cry. And cry they will, because sorting through all that data to find the needle in the noise will definitely not make them happier or give them the ability to find the bug.

What is needed in any monitoring system is the ability to increase or reduce the level of polling and data collection as system needs dictate. If you're actively debugging a system, then you probably want to turn the volume of data up to 11, but if the system is running well, you can dial the volume back down to 4 or 5. The volume can be thought of as the polling frequency times the amount of data being captured. Perhaps you want more frequent polling but less data per request, or perhaps you want more data for a broader picture but polled less frequently. These are the horizontal and vertical adjustments you should be able to make to your system at runtime. A one-size-fits-all monitoring system fits no one well. The fear, of course, is that by not having the volume at 11 you will miss something important, and that is a valid fear, but unless the whole reason for your existence is to capture all events at all times, you will have to find the right balance between 0 and maximum volume.

KV

2.17 I Don't Want Your Dirty PDFs

> *If you've got a bloodstain on your T-shirt, maybe dirty laundry isn't your biggest problem.*
>
> Jerry Seinfeld

For someone who cares a lot about kode, KV seems to write about documentation a lot, and there are pieces on documenting code elsewhere in this volume as well, including Sections 2.7 and 2.8. One of the many frustrations with documentation is the form in which it is packaged. While novels are probably best represented in a single, straightforward, narrative from which one rarely has to copy any information, technical documentation, such as data sheets and manuals, probably ought to be represented in an easy-to-consume form; one might even argue in favor of plain text!

The advent of systems such as *Markdown* should actually provide a useful middle ground between "plain ASCII" and something like PDF, and KV very much hopes to see more documentation, including chip manuals, provided in this format. Systems like Markdown have the advantage that they're close enough to plain text that they're easy to search with tools such as `grep` and edit in simple editors such as vi(m) and Emacs.

The next piece deals with what happens when people lock up their useful data in the somewhat misnamed Portable Document Format, which is portable to read, but not consume.

Dear KV,

I was recently implementing new code based on an existing specification. The documentation is 30 pages of tables with names and values. Unfortunately, this documentation was available only as a PDF, which meant that there was no easy, programmatic way of extracting the tables into code. What would have taken five minutes with a few scripts turned into a half-day of copy and paste, with all the errors that that implies. I know that many specifications such as Internet RFCs (Requests for Comments), for example, are still published in plain ASCII text, so I don't understand why anyone would publish a spec destined to be code in something as programmatically hard to work with as PDF.

Copied, Pasted, Spindled, Mutilated

Dear Mutilated,

Clearly you do not appreciate the beauty and majesty implicit in modern desktop publishing. The use of fonts (many, many fonts) and bold and underline increases the clarity of the written words as surely as if they had been etched in stone by a hand of flame.

Either you are facing a problem of overwhelming self-importance, as when certain people often in the marketing departments of large companies start sending e-mail with bold, underline, and colors, because they think that will make us pay attention, or you're dealing with someone who writes standards but never implements them. In either case this brings up a point about not only documentation, but also code.

Other than in mythology, standards and code are not written in stone. Any document that must change should be trackable in a change-tracking system such as a source-code control system, and this is as true for documents as it is for code that will be compiled. The advantage of keeping any document in a trackable and diffable format is that it is also possible to extract information from it using standard text-manipulation programs such as `sed`, `cut`, and `grep`, as well as with scripting languages such as Perl and Python. While it's true that other computer languages can also handle text manipulation, I find that many people doing what you suggest are doing it with Perl and Python.

I, too, would like to live in a world where when someone updated a software specification I could run a set of programs over the document, extract the values, and compare them with those that are extant in my code. It would both reduce errors and increase the speed with which updates could be made. The differences might still need to be checked visually, and I would check mine visually out of basic paranoia and mistrust, but this would be far easier than either printing a PDF file and going over it with a pen or using a PDF reader to do the same job electronically. While there are programs that will tear down PDFs for you, none is very good; the biggest problem is that if the authors had thought for 30 seconds before opening Word to write their spec, none would be necessary.

There are many things one can complain about with respect to the IETF (Internet Engineering Task Force), but the commitment to continue to publish its protocols in text format is not one of them. Alas, not everyone had Jon Postel to guide them through their first couple of thousand documents, but they can still learn from the example.

KV

2.18 Pining for a PIN

> *If someone steals your password, you can change it. But if someone steals your thumbprint, you can't get a new thumb. The failure modes are very different.*
>
> Bruce Schneier

The choosing of password lengths is a very common bike shed (see Section 5.2) among security practitioners, one that can be endlessly debated. With the advent of mobile computing devices, the arguments have continued to shift, and the addition of biometric mechanisms, such as fingerprint and face recognition, haven't really helped us to get any closer to one true answer. The problem is that there really isn't one true answer. Humans have crap memories, especially for long strings of unrelated digits or letters. Many koders, of course, find this lack of memory in normal humans strange, since many of us can easily remember all our credit card numbers and the phone numbers of our ten closest friends. Be that as it may, the typical user of technology cannot, and so we must work to come up with ways to help them overcome this problem.

Dear KV,

I'm converting a web application to run on a mobile platform. While the application doesn't handle banking information or anything with crazy security like that, it does still require the user to type a password. Our password requirements aren't too strict, though we do have a minimum size of eight characters and require one uppercase letter and one nonalphanumeric character. Because on-screen typing is so inaccurate, our product development folks want us to relax our password requirements even more, allowing the user to have a four-character, all-lowercase password, which they call a mobile PIN code. The mobile PIN would work only from the mobile app and not on the web. I've tried to explain to them that four characters simply aren't enough, but maybe if we're restricting this to the mobile app, it will be OK. What do you think?

Pinned Down

Dear Pinned,

I was wondering when someone would write a letter like this. Having typed on a variety of tablets in the past few years, I knew it was only a matter of time before some marketing or product-type person would ask an engineer to dumb down security for convenience. I'm glad you point out that your application doesn't relate to banking, as I would have had to write back immediately and ask, "Which bank?!" and that could only lead to trouble.

As a quick aside, I do find it interesting that the world thinks only of banks when they think of security. While it would be very bad for someone to transfer all the money out of your bank account into their own, the fact of the matter is that a lot of really bad things can happen online that don't relate directly to cash flow. Weak passwords can leave users open to identity theft, stalkers, kidnappers, and their exes. It's up to you to think about which of those is better or worse than losing some cash. I would take losing money over being stalked by my exes.

The problem with a four-character password, as you point out, is that it's too short and easy to guess. A PIN with a card, such as a card used at an ATM, is employed because it requires physical possession of something, the card, to be effective. Schemes for protecting users online, such as those involving passwords, depend on the user presenting a combination of something they have, something they are, or something they know. A password is an example of the latter. The schemes can be mixed and matched, such as the PIN and a physical card, where the user has something and knows something.

The problem with current mobile devices is that they are much more limited in their ability to handle user input than, say, a computer with a keyboard. On-screen keyboards aren't very good, which is probably why someone at Google thought to use a pattern for device unlocking on the Android. I find that system a bit silly and easy to shoulder surf, but it's good to see someone trying to do something differently.

It would be nice to pretend that the device itself would be proof of something the user has, but since the point of the password is to prevent a malicious party from accessing the user's data after the device has been stolen or, more likely, left in a bar, it needs to be sufficiently strong to deter an attacker, even if that attacker is also drunk. If you're unable to fight back on password complexity, it's time to break out the lockout option. A four-digit PIN code is mostly a problem if you allow the attacker a large number of attempts to guess the PIN. If, after three tries, you lock out the user for five minutes and then let the user try again, it's going to take a long time for the attacker to try enough PINs to guess the right one, that is, If the user hasn't picked a common PIN, such as 1234 or 2580 (an exercise for readers is figuring out why that second code is so common; for a more academic study of the problem of PINs, see the paper, "A birthday present every eleven wallets? The security of customer-chosen banking PINs," by Joseph Bonneau et al. from the Computer Laboratory of the University of Cambridge at https://www.cl.cam.ac.uk/~rja14/Papers/BPA12-FC-banking_pin_security.pdf).

Being the bastard that I am, I also like the idea of three failed tries causing the device not to work at all, including erasing all local data and then requiring the user to go through a recovery flow that does not involve the device. If your application is used by people who are often mentally impaired in some way (and no, I don't mean that they're upper management, but I do mean when it is used for social networking), then an annoying recovery system is not going to get past your product development folks. Too many people want to upload pictures of their friends in compromising positions, and that, alas, requires compromising on security.

KV

2.19 Reboot

When in doubt, reboot.

Anon

There is no better way to piss off someone who is trying to debug a problem than to remove the information they need to debug that problem, and yet so many people ask, "Have you tried unplugging the computer and plugging it back in again?" KV was recently asked this by a front-line tech who KV had asked about a problem in a web page. Instead of trying to explain how computers, web browsers, and the Internet actually work, all I could say was, "Seriously?" an interaction that has now been enshrined by several of my co-workers via screen shots of the aforementioned chat.

Nearly every system has some form of systems logging that it provides to programs; on Unix we have `syslog`, and all programs should be logging their data to such systems. There really is no excuse to not be logging pertinent information, but knowing what to log is an art in and of itself.

The art of debugging systems and software requires a lot of information, and doing what is described in the next letter is the absolute worst thing one can do; even if it is expedient to fix the symptom, it will never help to find the cure.

Dear KV,

One of my company's front-line engineers in the group that looks at the live traffic hitting our switches and servers keeps reporting problems, and then, before anyone can look at the server that's having issues, reboots the system to clear the problem. How do you explain to someone that there is information that needs to be collected when the system is misbehaving that is absolutely vital to finding and solving the problem?

Booted

Dear Booted,

I would start by standing with my foot on this person's chest and yelling, "There is information that needs to be collected when the system is misbehaving that is absolutely vital to finding and solving the problem." But I take it you've tried that already, though perhaps without enough screaming.

True, systems tend to build up state during execution that is not written to some permanent storage often enough. The problem you need to solve isn't preventing the person from insta-booting a misbehaving machine, as much as it is to make sure there is a good, searchable record of what the system is doing when it's running. Most system-monitoring tools on modern servers generate plain text output. It's a simple matter to write scripts that execute periodically to write the output of these tools such as `procstat`, `netstat`, `iostat`, and the like into files that will be preserved across reboots.

For more pernicious problems, you can write your own tools, either scripts or new programs that are executed when the system is shut down or rebooted. In this way, if people are insta-booting your machines before you can get to them, you can make it so that their reboot command does your bidding. You can even go so far as to rig your operating system to produce a kernel core dump on each reboot. This gives you a snapshot of the system as it was when it was broken, which you can go back to later and pick through. I warn you, though, that picking through a kernel core dump is about as much fun as picking fleas off a dog.

The only downside to collecting all this data is analyzing it. Since it's no longer really necessary to delete data, you may wind up spending a good deal of time organizing it into trees of trees of files. I offer a couple of quick suggestions. Do not make the tree scheme too difficult to traverse, either for a person or a program. It can take a very long time to access a ton of files in deep trees, due to the cost of traversing the directory trees. Keep things simple for both yourself and your analysis programs. Before you start, have a plan for what you want to store, where you want to store it, and how you plan to access it. Most people put this kind of thought into their applications, but not enough into how and where they store logs or other runtime information generated by their systems. You should put at least half as much time into the latter as you do into the former.

KV

2.20 Code Scanners

> *If you lie to the compiler, it will get its revenge.*
>
> Henry Spencer

The use of static analysis and other tools that are a bit like compilers on steroids to find potential problems in code has, for some, become standard practice, but these tools are still often looked upon with fear and derision because no one likes to have their errors pointed out to them. As recently as last month KV had to calm a group of developers about how these tools would be used. No, please try to imagine KV having to calm anyone, and you know how much fear these things still instill. The state of the art in static analysis still remains somewhat primitive with plenty of room left for improvement, and it's a shame that many of these tools are stand-alone and not integrated into the compiler. LLVM, the compiler du jour for people who want a good compiler, has started to integrate some features that were once only found in stand-alone static analysis packages, but more needs to be done here. Why bother having a large, post-processing step to tell the programmer of issues in their code if you can do it closer to the source, when they first wrote it, and the issues were fresh in their mind? It is for this reason that many people turn on compiler options that turn warnings into errors, which forces the koder to address lots more issues and prevents them from sweeping them under the rug. If there is one good use of fast build machines, it would be to apply these types of checks sooner, rather than later. Or, we can just use all that extra CPU time to mine idiotic coins.

Dear KV,

I'm in the QA group for a medium-size startup in Silicon Valley, and one of our VPs sits on the board of a company that makes code scanning software. You know, the stuff that spits out warnings about all the bad things you can do in C and C++. We've definitely found our share of buffer overflows and other problems in our code, but this stuff is expensive, over $5,000 a seat, and I'm just not sure it's worth it. What do you think of these tools?

Scanning for an Answer

Dear Scanning,

Over the years there have been both free and for-pay programs that would try to point out potential problems in people's code. Though the state of the art has certainly advanced since the `lint` program was written, there has consistently been one failing in the use of such programs, and that is the programmers themselves.

In the same way in which people build systems to log data and then never look at the logs, most people buy expensive and fancy tools to make their code faster, better, or more secure, only to ignore the advice that the tools give. I was recently working with an engineer on a project that was using an open source tool to scan his team's source code for possible buffer overflows and other security vulnerabilities. At one point the tool pointed out over 100 possible flaws in the system. The flaws were all identical and, it turned out, were false positives. Instead of using that age-old programming paradigm, the function call, to encapsulate the offending code, and thereby reduce the number of false positives from an annoying 100 to a tolerable 1, the engineer asked if we could "fix" the scanning tool so that it would not flag these false positives. It is at such moments that I'm happy to be bald, because if I weren't I'd be able to tear my hair out, and that hurts, or at least it used to. The impulse to "fix" the tool when it is giving plenty of good advice on other areas, and not to fix the code, or, more importantly fix the poor practices that lead to software problems, is going to be the biggest problem you will have.

So, my answer to you is "it depends." That is, it depends on what type of shop you work in. Do you work in a shop where people will see such tools as a help or as a threat? If people fight the tools, then they are worthless. If your programmers are willing to learn from the tools and to change their behavior, then the tools might be worth what you will pay for them. If your people need behavior modification, then I suggest you look to the film *A Clockwork Orange* for guidance. Just remember, Beethoven never coded a buffer overflow.

KV

2.21 Debugging Hardware

> *There has never been an unexpectedly short debugging period in the history of computers.*
>
> Steven Levy

Ah, hardware, the bane of every software developer's existence. If it weren't for this messy stuff dreamed up by diabolical electrical engineers, all our systems would perform exactly as we designed them. As someone who works on the lowest levels of software systems, KV has seen his share of awful software and awful hardware, but, alas, not the last of either or both. For most software developers the idea of trying to figure out misbehaving hardware ranges from horror to boredom, though, honestly, it's mostly horror. This letter tries to give some poor koder at least a few pointers in the right direction on how they should approach debugging a piece of hardware, rather than, as many of would prefer, defenestrating the offensive junk.

Dear KV,

What is the proper way to debug malfunctioning hardware?

Hard Up Against a Bug

Dear Hard Up,

I suggest taking a very sharp knife and cutting the board traces at random until the thing either works or smells funny! I gather you're not asking the same question that led me to use the word "changeineer" in another column ("Permanence and Change," *Communications of the ACM*, December 2008). I figure you have an actually malfunctioning piece of hardware and that you've already sent three previous versions back to the manufacturer, complete with nasty letters containing veiled references to legal action should they continue to send you broken products.

Along with race conditions, a subject for another time, hardware problems are probably the most difficult things to figure out. While hardware engineers may scoff at software engineers with screwdrivers, if you want to make them truly afraid, get out a logic analyzer or a scope and hook it up to their board. Most software engineers are not, alas, trained in using logic analyzers or even in basic electrical engineering, so you will have to content yourself with poking at the board through whatever software the board vendor or operating-system vendor has provided you.

Believe it or not (and I am sure if you're a typical software engineer you won't want to hear this), the best place to start is with the hardware vendor's documentation. Of course, many hardware vendors take as dim a view of documentation as software vendors do. The quality of the documentation I have seen has run the gamut from unusably terrible all the way up to "bang my head on the desk and cry." Rarely have I seen hardware documentation that was both correct and had a structure that made sense to anyone but the people who originally put it together. Happily, it is rare these days to be able to completely destroy a piece of hardware by putting the wrong value into the wrong memory location; the days of exploding computers à la the original "Star Trek" are still a couple of centuries in the future.

That being said, it is definitely possible to cause damage to hardware via software, or, more commonly, to mask whatever problem you were having by tripping some seemingly unrelated bit of configuration magic in the device. Not that KV is against magic; it's just that he tends not to trust it...at all.

If you're lucky, you have the documentation for the system, or can get the lawyer where you work to send a nondisclosure agreement and a letter to the vendor to get whatever it's willing to give you.

Read the documentation first. Really, trust me on this. It may be completely useless in the end, but it may also save you a lot of time if you find just the right bit of information in the docs. I tend to read over all the available registers and configuration options, of which there are often hundreds, and mark the ones I think might be related to my bug. I then tweak them one by one until I get a result. While this is a tedious process, it has been the one I've seen that has worked best.

Often you will not have a good way to interact with the hardware other than an already malfunctioning device driver. As devices have become more complex, vendors have released test and configuration programs that can be used to talk directly to the device, for example, over the PCI bus. If your hardware has such a program, and it works, then you are truly blessed. If, on the other hand, it does not come with such a program, there is a set of tools you can use to debug PCI-based devices, PCI Utilities, described at the end of this letter.

PCI Utilities have been ported to several operating systems, and something similar may exist in Windows, but, happily, that is not a form of pain to which I have been subjected.

If none of these yields results and you still have to "just get the thing working," it's time, alas, to call for help. The quality of the help you can get from a vendor seems to be linearly related to the price of the device. A cheap device usually comes from a low-cost producer who does not have the money to keep high-quality engineers on hand to help with problems, whereas an expensive device is more likely, but by no means guaranteed, to be produced by a company with experienced engineers. If you're specifying a device for a project at work, pick the one from the company that seems to have the better engineers. All devices have problems, but the ones that get fixed are the ones that have good engineering resources behind them. Cheap goods are cheap goods, in the end.

Once you reach a field or customer-support engineer, you need to be nice to them. I know, you're thinking, "What have you done with Kode Vicious?" but its true. Screaming at people and telling them they are idiots because they didn't consider your personal corner case is not the way to get your bug fixed quickly, even if you work for a large corporation and you have your CEO calling their CEO every day for a fix. You will need to work with this person or these people at least for the duration of your bug, so it's important to deal with them politely and professionally. Go back and read that again; I'll wait.

Lastly, you need to take good notes on the problem. There is nothing that is more frustrating than a bug report that says, "It's busted," and don't dare laugh, I've seen more than a few bug reports that say pretty much that. You need to be able to say how it is busted; when it was busted; if it stays busted, how to get it into the busted state; and any other information that seems related to the bug you're seeing. You should take notes not only on the bug but also on the fix. As you work with the engineers from your vendor, you need to track the patches they give you, if any; version changes in the hardware or driver; various theories about what might be wrong and whether the theories pan out; and pretty much everything else that is related to fixing or working around the bug. At this point you will often be both the project manager of the bug fix, as well as the remote hands for the vendor's engineers. While this may not be what you thought you signed up for, it's more often than not part of solving a hardware problem.

I hope you're lucky enough to have decent documentation and support from your vendor. If not, then I'll see you at the bar. I'm the guy sitting alone at the far end, crying into a chip manual with an always-full gin and tonic. My bartender knows me well.

KV

PCI Utilities

The PCI Utilities package contains various utilities dealing with the PCI bus, as well as a library for portable access to PCI configuration registers. It includes lspci for listing all PCI devices (very useful for debugging of both kernel and device drivers) and setpci for manual configuration of PCI devices (`http://www.linuxfromscratch.org/blfs/view/svn/general/pciutils.html`).

2.22 Sanity vs. Visibility

> *It is allowed on all hands, that the primitive way of breaking eggs, before we eat them, was upon the larger end; but his present majesty's grandfather, while he was a boy, going to eat an egg, and breaking it according to the ancient practice, happened to cut one of his fingers. Whereupon the emperor his father published an edict, commanding all his subjects, upon great penalties, to break the smaller end of their eggs.*
>
> Jonathan Swift,
> *Gulliver's Travels*

There is no koder working today who has not been touched by the battles of code formatting, and the continual war known as "Tabs vs. Spaces." It is such an old trope that it has now been part of a popular TV program, *Silicon Valley*. The age-old question has, of course, come to KV as well, and this was my response.

Dear KV,

My team resurrected some old Python code and brought it up to version 3. The process was made worse by the new restriction of not mixing tabs and spaces in the source code. An automatic cleanup that allowed the code to execute by replacing the tabs with spaces caused a lot of havoc with the comments at the ends of lines. Why does anyone make a language in which white space matters this much?

White Out

Dear White,

Ever edited a makefile? Although there is a long tradition of the significant use of white space in programming languages, all traditions need to change. We no longer sacrifice virgins to make the printer work, or at least KV hasn't sacrificed one recently.

In Python, many people have taken issue with the choice to have white space—and not braces—to indicate the limits of blocks of code, but since the developers didn't change their minds on this with version 3 of Python, I suspect we're all stuck with it for quite a bit longer, and I am quite sure that there will be other languages, big and small, where white space remains significant.

If I could change one thing in the minds of all programming language designers, it would be to impress upon them—forcefully—the idea that anything that is significant to the syntactic or structural meaning of a program must be easily visible to the human reader, as well as easily understood by the systems with which koders write kode.

Let's deal with that last point first. Making it easy for tools to understand the structure of software is one of the keys to having tools that help programmers prepare proper programs for computers. Since the earliest days of software development, programmers have tried to build tools that show them—before the inevitable edit-compile-test-fail-edit endless loop—where there might be issues in the program text. Code editors have added colorization, syntax highlighting, folding, and a host of other features in a desperate, and some might say fruitless, attempt to improve the productivity of programmers.

When a new language comes along, it is important for these signifiers in the code to be used consistently; otherwise your editor of choice has little or no ability to deploy these helpful hints to improve productivity. Allowing any two symbols to represent the same concept, for example, is a definite no-no. Imagine if you could have two types of braces to delineate blocks of code, just because two different parts of the programming community wanted them, or if there were multiple syntactic ways to dereference a variable. The basic idea is that there must be one clear way to do each thing that a language must do, both for human understanding and for the sanity of editor developers. Thus, the use of invisible, or near-invisible, markings in code, especially tabs and spaces, to indicate structure or syntax.

Invisible and near-invisible markings bring us to the human part of the problem—not that code editor authors aren't human, but most of us will not write new editors, though all of us will use editors. As we all know, once upon a time computers had small memories and the difference between a tab, which is a single byte, and a corresponding number of spaces (8) could be a significant difference between the size of source code stored on a precious disk, and also transferred, over whatever primitive and slow bus, from storage into memory.

Changing the coding standard from eight spaces to four might improve things, but let's face it, none of this has mattered for several decades. Now, the only reason for the use of these invisible markings is to clearly represent the scope of a piece of code relative to the pieces of code around it.

In point of fact, it would be better to pick a single character that is not a tab and not a space and not normally used in a program—for example, Unicode code point U+1F4A9—and to use that as the universal indentation character. Editors are then free to indent code in any consistent way based on the user's preferences. The user can have any number of blank characters used per indent character—8, 4, 2, some prime number, whatever they like—and programmers will have their very own personal views of the scope that they like. On disk, this format would incur only one character (two bytes) per indent, and if you wanted to see the indent characters, a common feature of modern editors is, you flip a switch, and voila, there they all are. Everyone would be happy, and finally we would have solved the age-old conundrum of tabs vs. spaces.

KV

3

Computer Science is the study of what can be automated.

Donald E. Knuth

Systems Design

Designing a system is fundamentally a different task from designing or implementing a single function. Developing a concise function can, at best, be a linear process. You consider the inputs, the expected outputs, and the possible errors, and if you're very lucky, you can stub it out and write it in a relatively linear fashion, working from the inputs to the outputs. While this might seem an oversimplification of the process of writing code, it's not too far from the truth for most koders, and it is definitely not the process most people would describe when designing any system of significant scale.

Well-designed systems depend heavily on one of the tenets described at the start of Chapter 3, which is composability. Just as good code is composable, a good system is made up of parts, be they functions, classes, modules, or whole programs that are themselves composable. The early developers of Unix put forth a systems design philosophy that has been variously quoted, but KV prefers Peter H. Salus's description.[1]

1. Write programs that do one thing and do it well.
2. Write programs to work together.
3. Write programs to handle text streams, because that is a universal interface.

The term *programs* comes from the fact that Unix programmers were often interested in building up many programs that could cooperate in processing streams of text, but we can apply the same advice to modules, classes, methods, and functions. The third suggestion has fallen by the wayside as systems have become more distributed and tend to

---

1. Peter H. Salus: *A Quarter Century of Unix*.

communicate more than just plain ASCII text, which was the norm when Unix was first built. Now it ought to be stated as, "Write programs with well-defined interfaces so that the output of one can easily become the input of another."

When designing a system, it is important to try to collect together all of the component parts into a consistent whole, one that describes, on different levels, the interconnections and dependencies among these parts. Two approaches to systems design have been promulgated in software development: top down and bottom up, and both have their proponents and detractors, much as the Lilliputians argued about which side to break their hard-boiled eggs. Very briefly, top-down design emphasizes having near total knowledge of the final system before work begins, while bottom-up emphasizes building up larger, complex systems, from pre-existing parts. Modern systems design nearly always comprises both approaches as very little software is written *from scratch* but is most often built up out of extant code, libraries, and other systems. The number of *green field* software projects built every year probably number less than 100, perhaps fewer than 10. For modern systems design we usually take quite a bit for granted, including tools like compilers, and platforms like the operating system and its supporting libraries, as well as a whole host of open source and proprietary components that need to be successfully sewn, some would say cobbled, together to make a coherent working whole.

We started out with our nose to the kode in Chapter 1 and pulled back a bit to look at Koding Konundrums in Chapter 2; now we try to pull back a bit further, just high enough above the code to be able to see how all the parts interconnect, in order to make a, hopefully, coherent system.

3.1 Abstractions

> *The art of programming is the art of organizing complexity, of mastering multitude and avoiding its bastard chaos as effectively as possible.*
>
> E. W. Dijkstra

Proper abstractions are the key to good systems design because if the abstractions are wrong, it's very likely that composability will suffer, and then the whole system will either not come together, or if it does, it will have warts, and pieces that seem like they don't fit together naturally. Failure to get the abstractions right leads to misunderstandings between components, which leads to bugs, which usually leads to system failure. When systems fail because the abstractions have not been thought through properly, the blame likely falls to the designer rather than the koders who had to implement the component parts.

Nearly every attempt to solve *the software crisis*, which was first identified in the 1960s and which goes on to this very day, has some relationship to the idea that if we can just come up with the right abstraction for what we want the system to do, all of our work will fall into place, and everything will be just fine. The fact that these abstractions continue to elude us after nearly 60 years does not seem to have caused a pause in the fads that periodically overtake the software industry. The plain old boring fact is that some level of abstraction is necessary and helpful in designing complex systems, but that too much of a good thing is counterproductive.

A good abstraction gives us a way to encapsulate an algorithm, in whole or in part, in a way that is usable, testable, and maintainable. Seems simple enough that it can be stated in a three-word list, and yet the particulars of each remain a challenge. Over the years we've developed the idea of functions, then libraries of functions, then modules, then came objects and object-oriented programming, all of which are abstractions that are supposed to help with reuse of the code we're already working on. Lists, tables, and trees of various sorts are also abstractions targeted at data, rather than code, and these too have proliferated like rabbits in Australia, with too much food and too few predators.

The central issue with various types of abstraction is not whether the concept of collecting related functions or data together is good or bad, or even, as in the case of object-oriented programming, keeping functions with the data they manipulate, is a good idea, as it probably is, but it is when the drive towards abstraction results in a Xeno's paradox of carving whatever we're working on into smaller and smaller bits until any particular bit, on its own, is relatively useless. Systems built only out of the smallest, understandable codelettes may be emotionally satisfying to a certain type of koder, but they generally result in an overly complex morass of code where it's nearly impossible to understand which part is doing productive work and which part is just there to glue the other tiny bits into the incohesive whole. The goal of using abstractions should be to reduce, and not increase, complexity. A system where there are too many tiny functions or methods effectively pushes the program logic into the connections between the functions; the callgraph defines most of the logic rather than the code in any particular function, which is classic spaghetti code. Such systems also have very high framework overheads, which waste memory and CPU in deference to someone's personal concept of elegance.

When we're looking at any abstraction, code or data, we must answer the three questions laid out above: Can this abstraction be utilized by a programmer as is? Can this abstraction be tested on its own? When it comes time to perform maintenance, how many knock-on effects do we have to worry about in the code that consumes this abstraction?

The utility of an abstraction can be gauged in two ways, proliferation of use and simplicity. A simple abstraction does not mean that it has a single operation, for example the plus operator, but that the operations that the abstraction provides are easy for the programmer to keep in mind when using it. Take, for example, the traditional file operations in Unix systems: `open()`, `close()`, `read()`, `write()`, `seek()`, and `ioctl()`. The file operations are a clear example of a good abstraction because they are related and easy to understand for the consumer, and they have, over time, been used successfully in millions of systems.

Our second measure, testability, can be related to simplicity, since it's far easier to test a piece of code, or a data structure, that has a small operating surface. A module with 10 operations is far easier to test than one with 100, no matter how good your test framework is, for it's not the framework that's being taxed, but the programmer's mind.

Maintenance is our final measure of an abstraction's quality. If I fix a bug in the code or change the layout of the underlying data structure, will that significantly change what happens with the existing code? How much re-testing is necessary when the abstraction is changed? A common problem is improving the speed of an abstraction, and who doesn't like more speed, and having that improvement break the assumptions of the code that consumes it. A change like this shows not that the consuming code was wrong, necessarily, but that there are assumptions between the consumer and the abstraction that were poorly understood, and, if this kind of problem recurs every time the abstraction is updated, then there is actually a problem in the abstraction, or the interface that it is providing to its consuming code.

Given how much of our time we spend dealing with abstraction, it is not surprising that there are many letters to KV about this very topic.

Dear KV

I have an office-mate who writes methods that are a thousand lines long and claims they are easier to understand than if they were broken down into a set of smaller set of methods. How can we convince this person that his code is a maintenance nightmare?

Fond-of-Abstractions

Dear FoA,

The short answer to your question is to make your office-mate maintain his own code forever, as that should be punishment enough. At some point they will realize that what they wrote three months ago is unreadable and begin to change their ways. Unfortunately, people with annoying habits, like talking loudly on cell phones, driving poorly, or giving unwanted advice, rarely see the errors of their ways. This is why there must always be crusaders for the good and right, such as ourselves.

I note that you used the word "methods" in your letter, rather than "functions," which indicates to me that you are using some form of object-oriented language. Before we go further let me point out that everything I say in this letter applies both to methods in object-oriented languages and to functions in non-object-oriented languages. The base problem is that there is too much functionality crammed into one place. There are several arguments you can make to explain why such over-long methods are problematic.

The first argument that comes to mind is code reuse. One of the reasons to have methods or functions in a program is to capture the essence of a single idea or algorithm so that it may easily be reused by others. When a method grows to 1,000 lines, it usually becomes highly specialized to one job, and a job that is probably not needed that often. It is far better to break down the larger problem into smaller ones, some of which may be reused by other parts of the software. Another added advantage of smaller, reusable methods is that they can be used in the next project you work on. Reusable methods are a benefit to the koder, who now does not have to write as much code, and to the company they work for because they can now finish a project faster. This kind of work avoidance is one of my favorite reasons for anything I do. Why should I work harder than I have to?

Another argument is that over-long methods are just plain hard to read and understand. A method that is 1,000 lines long, when viewed in a window that displays 50 lines at a time, works out to 20 pages of code to work through. Now, I don't know about anyone else, but 20 pages of anything, a book, magazine, or code, is hard to digest and keep all in my brain at any one time. Understanding anything requires context, and that context ought to be local. Jumping from page 18 back to page 2 because that's where the variable fibble was last modified often causes me to lose my place. I wind up staring at page 2 thinking, "No, why was I here?" I get all glassy-eyed, stare into space, and occasionally begin to drool, which makes my co-workers very nervous.

Finally we come to your excellent and well-justified point about code maintenance. Clearly if something is hard to understand, as we just established in the last paragraph, then it will also be hard to maintain. A thousand-line method is clearly doing too many things at once. How do you find the bug in a method when it is doing the equivalent of balancing your checkbook, whistling the "Ode to Joy," and juggling chain saws, all at once? Compounding the problem of just understanding such a method there is the issue that the number of possible side effects in a piece of code goes up quite a bit with every line you add. Perhaps not exponentially, but certainly more than linearly.

There are a few ways to set your office-mate on the righteous path to klean kode, if not clean living. One way is to make this set of arguments to them and see how they respond. Sometimes people can actually be shown the error of their ways. Using a neutral third party's code as an example is a good way to avoid the "I'm a better programmer than you" pissing match, which rarely wins anyone over to your side. If rational argument fails, you can try to use the software specification as a way to get reasonably sized chunks of functionality out of this person. You do have a specification for your software, right? If the specification clearly states the amount of work that is to be done for each method, then it will be pretty clear when this person violates the spec, and you can then come down on them as hard as you like at that point.

Of course, sometimes reasonable argumentation, i.e., the carrot, and direct control, i.e., the stick, fail. At that point I recommend something lingering, with boiling oil or melted lead. Just don't tell Amnesty International; though, if they had to maintain your office-mate's code, they might understand.

Fondly, *KV*

3.2 Driven

> *Data dominates. If you've chosen the right data structures and organized things well, the algorithms will almost always be self-evident. Data structures, not algorithms, are central to programming.*
>
> Rob Pike

Object-oriented systems in particular seem to suffer from an overwhelming amount of abstraction, which is unsurprising since a key selling point of object-oriented systems is that they allow abstractions to be made more obvious. Prior to object-oriented programming, abstractions had to be hand built, which could be done in any language, including such a low-level language as C.

Dear KV,

I've been working on a program in C++ to handle some simple data analysis at my company. The program should be a small project, but every time I start specifying the objects and methods it seems to grow to a huge size, both in the number of lines and in the size of the final program. I think the problem is that there are just too many things in the system to analyze, and each one needs a special case, which requires just a bit more code, or another sub-class to be created.

Help!

Driven to Abstraction

Dear DA,

One of the biggest problems when people use an object-oriented language is that when they realize how easy it is to create yet another class, they do. Instead of figuring out where the rubber meets the road, they instead find where the rubber meets the sky.

Of course, without looking at your code, and given your description, I really don't want to look at your code, I can't give you a pat answer. I also charge heavily for pat answers.

When I find someone I work with spending days specifying class after class without writing any implementation code, I tend to first take a long walk around the building. My therapist says that screaming at people helps no one. I don't agree with him, but for now, I am trying to play along.

I have a few pieces of advice when you find something you think should be simple starts to take up a huge amount of time and space. The first suggestion is to switch over from a compiled language, like C++, to something interpreted, like BASIC. Oh, wait, sorry, not BASIC. I meant Python, my current scripting language of choice. The reason I suggest Python is because it, too, is object-oriented, and it's easier to move an idea built in one OO language to another. You may even find that Python suits your needs perfectly and you won't have to move to a compiled language, but that decision is further off.

I suggest using a scripted language as my second piece of advice. Try to solve a smaller part of the problem you're working on. Programmers and engineers often try to bite off more than they can chew. We're a strangely optimistic lot, unless we're talking to a marketing person. In which case solving an equation such as 2 + 2 seems to require millions of dollars in investment, a colo full of machines, high-speed network links to everyone's house, and six weeks of paid vacation in Barcelona if you come up with a correct result. OK, maybe you don't handle your marketing department that way, but I can't suggest it strongly enough.

With a scripting language you can take smaller bites of the problem and play with them. If you can solve a segment of the problem and get some output to work with, you can then probably figure out the five or six next things to do and do them and so on.

The nice thing about working in smaller chunks is that you wind up with a result a lot sooner, and that's a lot more satisfying than having reams of UML diagrams and hand waving and a promise of a brave new world when you're done, which, at the rate you're going, you probably never will be.

So, get your tires out of the clouds, put them on the road, and implement a few things, instead of trying to solve everything at once.

KV

3.3 Driven Revisited

> *A program is never less than 90 percent complete, and never more than 95 percent complete.*
>
> Terry Baker

Abstraction also proved to be a popular topic with KV's readership, as the previous letter and response brought in three new responses, all of which are addressed in this section.

Dear KV,

I like your column on Koding in ACM Queue, mostly because at the end it says that you are an "avid bicyclist." Just like me. Bicycling in California, where you live, must be so much fun compared to Germany, where I live, because of the constantly nice weather you enjoy in California. Too much sun, however, can sometimes make you write columns that seem strange for us Old Europeans. In your column in Vol.3, No.2 ("KV Reloaded"), your advice to poor reader "Driven to Abstraction" is "to start with smaller parts of the problem" and to play with them using a scripting language.

Not only does California give you plenty of sun, it also has employers that give you plenty of time to play around with the "smaller problems" that you like in some programming language that is irrelevant for the later implementation. I always thought that in system design it may be convenient, but it is actually very dangerous to solve your favorite problem first and add the difficult, not-liked stuff later. Examples: Security—can it be added later? No! Read the pleas of guilty in "Patching the Enterprise"! Performance—can it be added later? No! Look into the faces of all the frustrated software re-engineers. Even more frustrating than the lack of sunshine in my country is the absence of employers that give you time to play around with scripting languages when project deadlines impend and budgets are already overspent. We are forced to make good overall designs first and improve it iteratively. Oh, California—the land of milk and honey and bicycling in the sun!

Koder-User-Rider-Teacher (KURT) ;-)

Dear KV,

Regarding the comment to switch from C++ to Python for the gentleman doing the "simple" data analysis. Two problems with question/answer scenario presented:

1) He doesn't understand the problem he's assigned, as he stated confusingly at the end of the paragraph that "there are just too many things in the system and each one needs a special case." Thus, it's *not* simple. He has a poor understanding of the problem assigned him. Statements like that should have come before he wrote any code at all. Had he spent more time mapping out the problem he could have saved himself some embarrassment.

2) You however did an equal disservice by blindly suggesting he switch to another language. How can you answer like that without knowing the details of the department? The solution, if nothing else, is to teach them how to think and design. This guy sounds like he was working in the boiler room with no one else. I remind you that you are writing for a prestigious international association, an association of professional "thinkers."

The *first* suggestion is for the developer to do the upfront thought and design needed so he understands the problem. An interpretive language is no better if he does not understand what the problem is, namely, the characteristics of the data involved and how it needs to be analyzed. Any engineer who "bites off more than he can chew" didn't do any engineering; he hacked. And you feed the crowd by suggesting nothing more than to hack with a different language. The real problem is not about what language to use but the lack of thought prior to writing code. But if my guess is correct, teams and managers don't have time for that. You simply suggested code as you go in the second to last paragraph.

I suggest the IEEE/EAI Standard 12207.0 known as the Development Process.

Sincerely, Karl Henning

My Dear Olde Europeans,

I want to thank you both for writing to KV here in the New World where we enjoy plenty of sunshine and benevolent employers. I was actually just enjoying a rubdown from my private masseur here at the office when your mails arrived, but I sent Jacques away so that I could concentrate fully on answering your letters.

What I believe you, unfortunately, missed in my original response was my suggestion that "Driven To Abstraction" break the problem down into smaller, possibly bite-size chunks. Although it is nice to think that you could specify all aspects of a program up front, that is only the case if all of the parts of the problem are understood before you start designing. I'm sure that each of you has been confronted with a system that you did not completely understand and that to be able to get a handle on the issues you had to work with smaller models and prototypes to be able to get your mind around the problem and finally solve it.

It is as easy to waste huge amounts of time over-specifying a system as it is to waste time playing with a scripted prototype of a subproblem, and each is equally dangerous to the success of a project. What is most dangerous, though, is simply staring at the same blank screen, page of your notebook, or white board, day after day without making any real progress. Telling your boss, "Well, I spent the last week thinking about the problem" is not acceptable, even here in the Land of Milk and Honey.

My suggestion was meant to break the mental stalemate that Abstraction had gotten into, the Zen equivalent of a tweak on the nose or a tap with a stick. I figure telling people to break down larger problems into smaller problems is more acceptable in the workplace than wandering around tweaking their noses or hitting them with sticks.

KV

Dear KV,

While I agree it's a good idea to not bite off more than you can chew, I'm concerned with your response to Driven to Abstraction's question. My fear is that your faithful readers will think it's OK to not create classes. I've seen too many supposedly-OO programs composed of a single class. I realize you don't suggest not adding classes, but you also do not directly address Driven to Abstraction's fear of classes.

Regards, *Afraid of Those Afraid of Classes*

Dear ATAC,

It's nice to see another take on the problem with my response to "Driven To Abstraction," and with this one I can wholeheartedly agree. Just two days ago I was reviewing some code that was clearly written by someone either afraid of or totally ignorant of classes. Actually, they seemed to also be ignorant of the concept of modularity, as everything was in a single, 4,000-line file. The kode was very clever, but in its current state it was totally unreusable. Unfortunately, this is a common problem I suspect we all face. Either due to time pressure or lack of training, someone decides to not only bite off more than they can chew, but more than anyone else can swallow.

As with many things in the world there is a spectrum between too big and too small. Too often people kode only for themselves without realizing that everything they create must be read and debugged by others. If we could only mend our selfish ways, perhaps we could all just get along.

KV

3.4 Changative Changes

> *Nothing is so painful to the human mind as a great and sudden change.*
>
> Mary Wollstonecraft Shelley,
> *Frankenstein*

As software systems grow and proliferate, the probability that an update to one library or component will break another becomes ever more likely, and since many components are loaded at runtime, both in operating system kernels and in applications, it becomes less likely that these issues will be found by the compiler, linker, or anywhere else in the build system. Attempts to solve this problem are usually encoded into package systems, which attempt to resolve these conflicts when a package is updated, by keeping track of all of the dependencies and forcing an update of all of the relevant components to a, hopefully, compatible version. Current package systems do this not by understanding the components at the API level but at the level of the overall version, which is insufficiently granular and also prone to errors, because the dependencies are expressed by human beings marking versions of libraries as compatible or incompatible. One proposal is to be stricter about what the version numbers mean, with the major number increasing only when a completely incompatible change is made, the minor number changes when the code is no longer backwards compatible, and the patch number, the last digit, is increased for each patch or minor change. Giving a more concrete meaning to version numbers won't solve the problem of depending on humans, but having such a standard would make it easier for koders to know if they're about to eat a change that might give their program indigestion.

Dependency analysis is an area ripe for automation, since, for compiled language, a signature could be generated for every function entry point, based upon its name as well as the names and types of its arguments and return value. A change to the name, as we see in the following letter, is easy, but changing the types of arguments or the return value is often missed by languages that are looser in their interpretation of types. Compilers already record plenty of data about function entry points, as these are necessary for the debugger, and therefore an extension of this mechanism to aid packaging and dynamic loading systems is most welcome, and not just one that throws an oblique error about incompatibility, but that explains just which thing is incompatible, down to the specific entry point.

Here we see how disastrously this can go wrong, and, unfortunately, this is a very common failure still seen in software systems.

Dear KV,

For the last two years I've been working on a software team that produces an end-user application on several different operating system platforms. I started out as the build engineer, setting up the build system, then the nightly test scripts, and now I work on several of the components themselves as well as maintaining the build system. The biggest problem I've seen in building software is the seeming lack of API stability in software. It's OK when new APIs are added, you can ignore those if you like, and when APIs are removed, you know because the build breaks. The biggest problem is when someone changes the API. This isn't discovered until some test code, or worse, a user executes the code and it blows up. How do you deal with constantly changing APIs?

Changes

Dear Changes,

The best way to deal with change is to bury your head in the sand and ignore it. After all, we can all learn from the great management traditions of the past, and engineers are no exception to this. Hmm, perhaps not.

What you point out is one of the biggest challenges in building large and complex systems. Software is amazingly malleable, and that makes it possible, and unfortunately quite probable, that someone will make a change. What many engineers and programmers don't realize is that when they're building a library, or really any component that others are supposed to depend on, the API becomes the contract between their code and everyone who uses it.

As you point out in your letter, there really three ways in which this happens. The first, adding an API, won't affect your system because with no one to call it the new API can't really cause much damage. The second case, removing an API, results in an immediate error when your program is linked, either at compilation or run time, so at least you notice this before trying to really use the code. The last case is the one that will give you fits and nightmares because there are very few automated ways of finding an API that looks the same, but isn't. At one place I worked we dubbed this as "changative change" for want of a better phrase, or, it would seem, a technical writer.

On one particular system about 80% of our problems were related to trying to re-integrate different subsystems with each other. The problem, as you can imagine, grows quite quickly with the number of components involved. Two subsystems that depend on each other have at least one dependency, whereas 4 subsystems have 6 dependencies, and 8 sub-systems have 28, and so on. Building up any sort of coherent system from a set of modules, all of which are changing, turns out to be very hard, but there are some solutions.

Operating systems people have long known about this problem, and so APIs that programs depend upon tend to change only very slowly, or not at all. The basic open( ), `close()`, `read()`, `write()` system calls in Unix and Unix-like operating systems have taken the same arguments and returned the same types of values for nigh on to 20-plus years at this point. When new subsystems were added, such as networking, new function calls were added as needed; hence, to open a socket, you don't call `open()`, because that would have required changing its arguments and therefore all the code that already used it. Instead, you have the `socket()` system call, which takes different arguments but returns a value that is usable by `read()` and `write()`. System programmers also tend to narrowly define the set of functions they will provide, because they know the nightmare of maintaining an arbitrarily wide set of APIs. FreeBSD, for example, has around 400 available system calls, that is, APIs that user programs are allowed to call to get the OS to do something for them, like read a file or find out the time. Although that number is not small, it is trackable and maintainable, whereas the number of APIs in the full set of POSIX libraries, or Microsoft Foundation Classes, is far, far larger.

Another trick that can be adopted from the systems programming world is that of the `ioctl()`, or I/O control. Device driver writers can do most of the necessary work using the simple `open()`, `close()`, `read()`, and `write()` semantics, because what most people want from a device is to open or use it, read data from and write data to it, and then put it away, or close it. The problem here is that it is often necessary to have device-specific controls that can be easily exported upwards to the operating system, for example, to set a network device into promiscuous listening mode, or to set its various address parameters. It is these special cases where `ioctl()` is used. The `ioctl()` call has been used, and verily, abused, over the years, but the basic design principle is a sound one. Always leave yourself an escape route.

Lastly, there is discipline, which some people like, but this is not that kind of book. What I actually mean is that there has to be a decision made about how changes get made in a system. Changing things fast seems to be in vogue at the moment; the so-called extreme programming methodology has led to a lot of this. Many engineers simply decide that at some point an API is set in stone and that it has too many callers to change, and so any changes require new APIs.

Unfortunately I doubt I've solved your real problem, because unless you and your team write everything from scratch you will be at the mercy of people who can, and will, make mistakes. My only other advice is that your team use the smallest number of external APIs possible and to not use too many new or advanced features as those are the ones that are mostly likely to change.

KV

3.5 Threading the Needle

*Why Threads Are A Bad Idea
(for most purposes)*

John Ousterhout

Sometimes a good quip will last for, and influence, a generation. In the area of programming, John Ousterhout's commentary on threaded programming, quoted above, is quite well-known and so it was unsurprising when a letter came to KV to ask for thoughts on this topic. The topic has come up more than once as it's also addressed in Section 3.6. Given the realities of modern hardware, which achieves performance by giving the programmer many cores on which to execute their code, it is impossible to not consider threaded programming in solving many significant software problems, which means everyone must now learn and understand how to write and debug threaded programs.

Interested readers can find the original presentation from Ousterhout here:

`https://web.stanford.edu/~ouster/cgi-bin/papers/threads.pdf`

Dear KV,

When I was in school, I read a paper on how threads were considered dangerous, but that was before most CPUs were multicore. Now it seems that threads are required to get improved performance. I haven't seen anything that indicates that threaded programming is any less dangerous than it used to be, so would you still consider threaded programming to be dangerous?

Hanging by a Thread

Dear Threaded,

You might just as well have asked me if guns are still dangerous, because the answer is closely related: only if the gun is loaded, and definitely if the business end is pointed at you.

Threads and threaded programming are dangerous for the same reasons they always were: because most people do not properly comprehend asynchronous behavior, nor do they do a good job of thinking about systems in which two or more processes work independently.

The most dangerous people are those who think that simply by taking a single-threaded program and making it multithreaded, the program will somehow, as if by magic, get faster. Like all charlatans, these people should be put in a sack and hit with a stick (an idea I got from the comedian Darragh O'Brien, who wants to use that method for psychics, astrologers, and priests). I'm just adding one more group to his list.

Probably my favorite example of not thinking clearly about threaded programming was a group that wanted to speed up a system they had developed that included a client and a server component. The system was already deployed, but when it was scaled up to handle more clients, the server, which could handle only one request at a time, couldn't serve as many clients as was called for. The solution, of course, was to multithread the server, which the team dutifully did. A thread pool was created, and each thread handled a single request and sent back an answer to a client. The new server was deployed, and more clients could now be served.

Just one thing was left out when the new server was multithreaded: the concept of a transaction identifier. In the original deployment, all of the requests were handled in a single-threaded manner, which meant that a reply to request N could not be processed before request N–1. Once the system was multithreaded, however, it was possible for a single client to issue multiple requests and for the replies to return out of order. A transaction ID would have allowed the client to match its requests to the replies, but this was not considered; and when the server was not under peak load, no problems occurred. The testing of the system did not expose the server to a peak load, so the problem was not noticed until the system had been completely deployed.

Unhappily, the system in question was serving banking information, which meant that a small but nonzero number of users wound up seeing not their own account information but that of other customers, resulting in not just the embarrassment of the development team, but the shutting down of their project, and in several cases, firings. Alas, the firings were not out of cannons, which I always felt was a pity.

What you ought to notice about this story is that it has nothing to do with inter-thread locking, which is what most people think of when they're told that a piece of code is multithreaded. There is no magic method to make a large and complex system work, threaded or not. The system must be understood in total, and the side effects of possible error states must be well-understood. Threaded programs and multicore processors don't make things more dangerous per se; they just increase the damage when you get it wrong.

KV

3.6 Threads Still Unsafe?

Unsafe at any speed.

Ralph Nader

The fight over threaded code seems as if it will never end. Threads are a key abstraction used in decomposing a system into cooperating parts, but our tools for understanding them remain woefully inadequate, to the point where newer computer languages, such as Go, have tried to make their use both explicit and comprehensible. The problem with threads isn't just with our tools, but also with our minds. Experience with software design has shown the number of people who can comprehend how to build a system out of cooperating, relatively uncoordinated, independent tasks, is relatively small.

Dear KV,

Due to performance needs, my team is re-working some old code to run multithreaded so that it can get some advantages from the new multicore CPUs that are now shipping in high-end servers. We're estimating that it will take at least six months to break down our software such that it will be granular enough to run as multiple threads and to implement all the proper locking and critical sections. I happened to come across an old paper online when looking up other information on threading, "Threads Considered Harmful," and was wondering what you thought of it. The paper was written long before we had multicore CPUs, and at the time there were few commercial SMP machines, so perhaps it didn't make sense to go to all the bother of writing threaded code then, but now, things are different. Have you heard of this paper? Do you think it is still valid?

Hanging by a...

Dear Hanging,

John Ousterhout's warning is as important today as it was when it was written, not because times and technology haven't changed, but because, alas, people haven't. Most people seem to decide to create multithreaded code for the reasons you state here, that is, because of wanting to get a supposed performance boost from it. These same people never seem to bother to measure their code or to see if it is even practical to run it in multiple threads; they just start slicing away at the code in the vain hope that if they have enough threads suddenly, as if by magic, their code will run faster.

Longtime readers of KV will know that I do not believe in magic bullets. Waving a wand labeled "threads" over your code is about as likely to make it run faster as is sacrificing a chicken. At least you can eat the chicken when you're done with it, which is more than I can say for your code. In actual fact threading your code may make it run slower, because poorly written threaded code is often slower than poorly written nonthreaded code. The locking primitives required to get locking right are nontrivial and can, if improperly used, slow your code down as it all blocks on the same lock, or, even worse, introduce subtle bugs.

Another problem with threaded code is that the tools used to debug it remain primitive. Although most debuggers now claim to handle threads properly, this is in fact not always the case, and you really don't want to be debugging your debugger while debugging your code. Race conditions are as hard to debug now as they were 20 years ago, and they don't seem to be getting any easier to find, let alone fix.

One last thing that a lot of people miss in their rush to thread their code is the support they get from libraries that they link against. If your program requires that the libraries it uses be multithreaded as well, then you may be in for a shock when you realize that some of them are not thread safe. Using non-thread-safe libraries in your thread-safe program is going to cause you no end of trouble.

Given all of this, should you still continue to go about threading your code? Maybe. First you need to understand the trade-offs and see if the job the code does is amenable to being multithreaded. If the code has several components that can operate completely independently, then, yes, multiple threads can be a boon; if, on the other hand, the components all need to access a small shared section of data all the time, then threads will get you nowhere. Your program will spend most of its time acquiring, freeing, and waiting on the locks that protect the shared data.

So, unless it's really a win and you and your team have thought about it a lot, I'd try not to get hung up in threads.

KV

3.7 Authentication vs. Encryption

Security is a state of mind.

NSA Security Manual

One would think that after 20+ years of having a public network on which people buy and sell various goods that most people who work with technology would understand the difference between authentication and encryption, but as this letter shows, that knowledge is not as pervasive as one might prefer. The online world might be a better place if these concepts were not only well understood, but also applied intelligently and consistently, to pretty much all software systems, and yet...

Dear KV,

We're building out a new web service where our users will be able to store and retrieve music in their web account so that they can listen to it anywhere they like, without having to buy a portable music player. They can listen to the music at home with a computer hooked to the Internet or on the road on their laptop. If they want, they can download music, but if they lose it through a problem with their computer, they can always get it back. Pretty neat, huh?

Now to my question. In the design meeting about this I suggested we just encrypt all the connections from the users to the web service because that would give us and them the most protection. One of the more senior folks just gave me this disgusted look, and I thought she was really going to lay into me. She said I should look up the difference between authentication and encryption. Then a couple of other folks in the meeting laughed, and we moved on to other parts of the system. I'm not building the security framework for the system, but I still want to know why she said this? All the security protocols I've looked at have authentication and encryption, so what's the big deal?

Sincere and Authentic

Dear Authentic,

Well, I'm glad she laughed, screaming hurts my ears when it's not me doing the screaming. I'm not sure what you've been reading about cryptography, but I bet it's some complex math book used in graduate classes on analysis of algorithms. Fascinating as NP completeness is, and it is fascinating, these sorts of books often spend too much time on the abstract math and not on the concrete realities of applying the theories in creating a secure service.

In short, authentication is the ability to verify that an entity, such as a person, a computer, or a program, is who they claim to be. When you write a check, the bank cashes it because you've signed the check. The signature is the mark of authenticity on that piece of paper. If there is a question later as to whether you actually wrote me a check for $1,000,000 dollars, let's say if I decide to deposit it in my bank account, then the bank will check the signature.

Encryption is the use of algorithms, whether they're implemented in a computer program or not, to take a message and scramble it so that only someone with the correct key to unlock the message can retrieve the original.

It's pretty clear from your description that authentication is more important to your web service than encryption at the moment. Why is this? Well, what you care most about in your situation is that users can only listen to the music they've purchase or stored on the server. The music does not need to be kept secret because it is unlikely that someone is going to steal the music by sniffing it from the network. What is more likely is that someone will try to log into someone else's account to listen to their music. In order for a user to prove who they are, they will authenticate themselves to your service, most likely via a username and password pair. When the user wants to listen to their latest purchase, they present the username and password to the system in order to get access to their music. There are many different ways to implement this, but the basic idea, that the user has to present some piece of information that identifies them to the system to get service, is what makes this authentication and not encryption.

The password need not be encrypted, only hashed, before being sent to the server. A hash is a one-way function that takes a set of data and transforms it uniquely into another piece of data from which the original cannot be retrieved by anyone, including the author of the hash function. It is important that the hash function produce unique data for each input, as collisions make it possible for two different passwords to be the same hashed data, and that would make it harder to differentiate users.

There are plenty of books and papers on this sort of stuff, but try to avoid the pie in the sky stuff unless you're researching new algorithms, because you really don't need it, and it'll just make your head hurt.

KV

3.8 Authentication Revisited

> *The trouble with quotes on the Internet is that they're very hard to authenticate.*
>
> Abraham Lincoln

Oftentimes the responses to KV's writing are even more interesting, and impassioned, than the letter that started things off. The next letter and response bring up an interesting side of the issue discussed in Section 3.7.

Hello dear KV,

Supposing I'm a customer of Sincere-and-Authentic's and suppose the sysadmin at my ISP is an unscrupulous, albeit music-loving, geek. He figured out that I have an account with Sincere-and-Authentic. He put in a filter in the access router to log all packets belonging to a session between me and S-and-A. He'd later mine the logs and retrieve the music. Without paying for it.

I know this is a far-fetched scenario, but if S-and-A want their business secured watertight, shouldn't they be contemplating addressing it too? Yes, of course, S-and-A will have to weigh the risk against the cost of mitigating it, and they may well decide to live with the risk, but I see your correspondent's suggestion as at least worth a summary debate, not something that should draw disgusting looks! There's in fact another advantage to encrypting the payload, assuming IPSec isn't being used: decryption will require special clients, and that will protect S-and-A that much more against the theft of their merchandise.

Balancing is the Best Defense

Dear Balancing,

Thank you for reading my column in the April 2005 issue of Queue. It's nice to know that someone is paying attention. Of course, if you had been paying closer attention, you'd have noticed that S&A had said, "In the design meeting about this I suggested we just encrypt all the connections from the users to the web service because that would give us and them the most protection." That phrase "just encrypt all the connections" is where the problem lies.

Your scenario is not so far-fetched, but S&A's suggestion of "encrypting all the connections" would not address the problem. Once the user had gotten the music without their evil ISP sniffing it, they would still be able to redistribute the music themselves. Or, the evil network admin would sign up for the service themselves and simply split the cost with, say, 10 of his music-loving friends thereby getting the goods at a hefty discount. So, what S&A really needs is what is now called Digital Rights Management. It's called this because for some reason we let the lawyers and the marketing people into the industry instead of doing with them what was suggested in Shakespeare's *Henry VI*.

What S&A failed to realize was that the biggest risk of loss of revenue was not in the network, where only a small percentage of people can play tricks as your ISP network administrator can, but at the distribution and reception points of the music. Someone who works for you walking off with your valuable information is far more likely than someone trying to sniff packets from the network. Since computers can make perfect copies of data, after all that's how we designed these things in the first place, it is the data itself that must be protected, from one end of the system to the other in order to keep from losing revenue.

All too often people do not consider the end-to-end design of their systems, and instead "just" try to fix one part.

KV

3.9 Authentication by Example

> *We should treat personal electronic data with the same care and respect as weapons-grade plutonium; it is dangerous, long-lasting, and once it has leaked, there's no getting it back.*
>
> Cory Doctorow

Now that we all know the difference between authentication and encryption we can turn our attention to the proper use of authentication. I'd like to believe that this letter is old enough that no one in their right mind would ever think of doing what the original author did to create their first pass at an authentication system, but, at this point, I think we all know better.

At a somewhat higher level there remain many issues in building and fielding an authentication system, beyond those that are covered in the letter. One of the key questions is longevity: How long should an authenticated session be allowed to last, and should continued use extend the time of a particular session?

The answers to this question run the gamut from terminating a session after a few minutes of idle time, in the case of banking applications, to seemingly forever for chat systems like Slack, or cat watching web sites like Facebook. The banking world has a default deny policy, which is not meant to protect their customers but to cover the bank's...potential losses, showing once again that the only way to get things done is to start a lawsuit. If you're not a banking application, how do you decide a proper session timeout? Is there some algorithm we can apply to make this decision easy? Yes, these questions are rhetorical.

Choosing a session timeout comes down to judging the downside risk of an attacker being able to acquire the same rights as a valid user of the system. If the only risk is that the attacker can also see pictures of cats, a read-only ability of, free, public, nonthreatening data, then the session timeout can be quite long. Consider the front page of an online newspaper. Most newspapers now have paywalls, with varying degrees of success, but they all leave their front pages open, or else how would they attract readers to their site? Such a system can of course have a long timeout, or, no timeout at all, so long as the only capability is the one of reading the free pages of the news. Once past the paywall, of course, the session needs to have some timeout, or else an attacker can get an infinite token with which to bypass, and therefore not pay for, the paywall. An infinite token could (and would) be posted to a system like reddit, or 8chan, for use by everyone who did a search for it. Now that we know we need a timeout, how long should it be? How often do people check the news? Probably for a news site the timeout should be a few hours, as this is how long we might expect someone to check their daily news, and maybe we extend it to nearly a day, so that the user has to log in again the next day, as if they were paying for a daily paper. The decision will, of course, have to be worked out with other groups, such as marketing, but I'll not go into how to deal with people like that here.

One might be tempted to think that the shortest timeout is the best, but this is actually not always the case, especially if a session timeout causes the user to type their password again. The more often a user has to type their password, the more likely they are to write it on a sticky note and stick it to their desk. While I wish this were a joke, it's not. Human beings have terrible memories for things like passwords or passphrases, and not even the advice of Randall Munroe seems to be able to help them; see https://xkcd.com/936/. With the advent of fingerprint and face recognition systems for phones, tablets, and some laptops, the password problem has been reduced, but not eliminated. Switching to having something that you are, rather than something that you know, can change the session timeout calculus, but I know of no web site, banking or otherwise, that does not use a password, even if biometrics are an option.

So now let's return to the letter, which isn't about session timeouts, but once you've read it and the response, you'll see that the session timeout question follows not long after.

Dear Kode Vicious,

I am a new webmaster of a (rather new) website in the company's intranet. Recently I noticed that although I have implemented some user authentication (a start *.asp page linked to an SQL server, having usernames and passwords), some of the users found out that it is also possible enter a rather longer URL to a specific web page within that web (instead of entering the web's homepage), and they go directly to that page without being authenticated (and without their login to be recorded in the SQL database). It makes me wonder what solution you could advise me to implement in order to ensure any and all web accesses to be checked and recorded by the web server.

New Web Master

Dear NWM,

You have my deepest sympathies; users, as I have noted before, are the bane of our existence. Those sneaky bastards will go around your systems every time just to get the data they want, without paying your login system any heed. Now, there are many ways to deal with recalcitrant users, but unfortunately I can only discuss the ones that are legal here; the others, trust me, are far more enjoyable.

Not to be too cruel, but it seems from your description that you have created a superfluous authentication system that doesn't provide much for the users or for you. Users can and do go around your login page, and therefore you've just created more code without any real value, and valueless code is a real shame. You need to change your way of thinking about this problem before you can solve it.

Right now using your authentication system is voluntary, and therefore easily bypassed, because you are not enforcing the authentication on each page that your users can see. In your letter you don't say that your system has any of the necessary features of an authentication system, such as:

- What the authentication gives the user the right to do. For example, can they read, modify, or create pages?
- How the user proves to the system that they were authenticated.
- How the user and the system agree that the user is who they say they are.

You have a system of web pages, which you believe contain valuable information since you claim you want to protect them. Yet those pages have no protection from anyone who can work out or guess what the name of the link is?! That's just plain wrong. If you have information that must be protected, then you should be protecting it. One way to protect it is to implement the features shown in the list, namely, that the user must prove who they are to your login system and then must prove they were authenticated and have a right to see the information whenever they want to read a page.

How does the user prove they have these rights? The user must first talk to the login system to prove who they are. In your case you implemented a web page that fronts a database of usernames and passwords.

As a quick aside, I hope you are storing only the hash of the password and not the raw text. A hash uniquely destroys the password so that if the password was *foo*, the resulting hashed value might be the number 5. Given the number 5 someone cannot get back to *foo*, but given *foo* and the same hash function, you will always get 5. When verifying a password, you're really comparing the hashed values; the original string *foo* is never stored. Keeping a database of raw username and password pairs is a serious security hole, the kind of bug that makes KV want to make a big pot of programmer hash.

So, now the user can prove who they are by logging in with their username and password, but how can you satisfy feature 3? They submit a username and password to your system, but, well, so what? The real problem with your pages is that the pages themselves are not protected. If you want to force users to authenticate themselves, then each request to your web server must include some piece of information that proves the user has been authenticated. If the server does not check requests to see if they come from authenticated users, then there is no point in having an authentication system at all. What should the user have to present to the system?

In the web world the most common piece of information exchanged between a user, or more exactly the user's web browser and a server, is a cookie. A cookie is just a chunk of data that can be set by a server into a user's web browser. When the user is browsing within a particular domain, the server can look at the cookie and get information from it. In an authentication system the server should set a cookie into the user's browser that is then checked on every subsequent access to the system to validate that the user has the right to use the system.

What should go into the cookie? Well, that's mostly up to the data that you, as the system maintainer, wish to track, but two things are absolutely necessary to prevent the system from being abused. The first is that the cookie must have a digital signature. A digital signature can prevent the cookie from being manipulated by a user to gain access when they should not. If someone were to find out the format of your cookies and your cookies were not signed, then that person could just craft their own cookies and present them to the server, bypassing your authentication system. The second necessary part is that the cookie should contain a timeout, past which the cookie is no longer good and must be replaced. Infinite timeouts on authentication cookies makes those cookies very valuable to steal, because once acquired they can never be revoked. Picking the timeout, though, is a balancing act. Your users will want their authentication tokens to last as long as possible, perhaps for months, while to maintain control of your systems you would prefer something much shorter, such as one hour. Finding the compromise between the permissive and the fascist timeout is beyond the scope of this article and depends on your users and how much management backup you have to force them to behave in the way that you'd like. I find that reminding management of how much money they'll lose if users are able to leak and leech information from the system to be very effective in getting shorter timeouts implemented. Management hates losing money like KV hates losing the keys to his liquor cabinet.

So, now you have a model of an authentication system, instead of just a system that happens to record users logging in when they feel like it. Users log in, they get a cookie that is digitally signed to prevent tampering and that has a limited lifetime, and they must then use that cookie to read any of the other pages. There are many ways to implement this, but that's the general outline, and there are plenty of examples of how to do this on the net, so get back in there and fix this!

KV

3.10 Cross-Site Scripting

> There is no such thing as perfect security, only varying levels of insecurity.
>
> Salman Rushdie

One of the crosses I have had to bear in preparing this volume is realizing how many times the same topic has arisen, and even though the advice given is simple and direct, the problem continues, nearly unabated. The following response on cross-site scripting was written more than a decade ago, and yet if I were to type those words into a search bar, or for even more fun, Mitre's Common Vulnerabilities and Exposures search (https://cve.mitre.org/cve/search_cve_list.html), I would find that not only does the problem continue unabated but that there are 341 total search results with 4 of them occurring in the first few months of this year.

The reason for nearly all cross-site scripting (CSS) vulnerabilities is a near complete disregard for properly validating input, which is the over-arching problem, while CSS is just a specific instance. I wish I could say that it's amazing to me that after 20+ years of people developing code that faces the global Internet, that koders still fail to validate user input, but then most people still don't wash their hands after they use the toilet, so maybe people just never learn.

The input validation meta issue brings us to three important points in system design when handling user input:

- Never pass user input to anything that will treat the input as something to execute.
- Try to match any user input against any known good pattern.
- Use the built-in sanitation routines in your chosen language.

Every language, for better or for worse, usually for worse, has a way to call out to the system on which it's running to get some work done, such as deleting a file, changing permissions, or starting another program. These catchall idioms usually look like the `system()` routine found in the C library, but nearly every language has them, Python, PHP, Go, Rust, and on and on. The best course of action is to absolutely never use this idiom. The second best course of action is to never allow input from a user to find its way to that idiom. The problems caused by this idiom are so common that nearly all static analyzers will search your code and put big screaming warnings up when they find such a thing.

Now that we're past the point of handing user input to the `system()` routine, let's think about how we might properly accept input from the user. While it is not always possible to anticipate every user input, there are many cases in which we are only interested in a few types of responses, such as a predetermined set of answers to a question. If we are lucky enough to be in this situation, then we can build that predetermined list into our system and disallow any input not part of the list. Revisiting Postel's early Internet programming wisdom, we would be conservative in what we accept.

Finally, for point three, we come to the case where we have to deal with nearly arbitrary input from the user, and this is the case that is covered in the following letter and response.

Dear KV,

I know you usually spend all your time deep in the bowels of systems with C and C++, at least that's what I gather from reading your columns so far, but I was wondering if you could help me with a problem in a language a little further removed from low-level bits and bytes, PHP. Most of the systems where I work are written in PHP and, as I bet you've already worked out, those systems are web sites. My most recent project is a merchant site that will also support user comments. Users will be able to submit reviews of products and merchants to the site. One of the things that our QA team keeps complaining about is possible XSS attacks, cross-site scripting. Our testers seem to have a special ability to find these and so I wanted to ask you about this. First, why is cross-site scripting such a big deal to them, second how can I avoid having such bugs in my code, and finally why is cross-site scripting abbreviated XSS instead of CSS?

Cross with Scripted Sites

Dear CSS,

First, let's get something straight, I may spend a lot of time with C and C++, but I object to the use of the word "bowels" in this context. My job is bad enough without having to now have this image of literally working in the bowels of anything.

Let me answer your last question first, since it's the easiest. The reason that cross-site scripting is abbreviated XSS is the same as the reason that I spell code as kode. Programmers and engineers think they're clever and like to put their mark on things by changing the language, in particular turning every possible term they coin into some acronym that only they know. It is one of the side effects of specialization that we will leave alone, just now, before my more literate friends come after me with torches and pitchforks.

Now, back to what I think we can both agree are your more serious queries: cross-site scripting, that is, the ability to inject JavaScript into a site and to then have the site send that scripting code on to the user. There are actually many risks involved in cross-site scripting attacks because the JavaScript code can do many different malicious things. For example, the code can completely rewrite the displayed HTML, which in your case means that someone else would be able to completely overwrite the reviews that the user submitted, probably not something you'd like others to be able to do. Another example is that the malicious code can steal the user's cookies, and cookies are often used in web applications to provide identify the user. If the user's cookies get stolen, then the attacker can become the user and perhaps take over their account. If your site uses cookies in this way, this is a pretty big risk. So, you can see why QA gets their knickers in a twist, and to be honest I'm surprised they never bothered to explain just why this was a risk, or maybe they just assumed you knew better.

Winding up with a cross-site scripting bug is almost always the result of not doing proper input validation. Since you say that you've read earlier columns, then you must know that I don't trust users, and neither should you. When designing a web site, you have to just accept that with millions of potential users some percentage of those people who use your site will attack it. It's the way the world is, some people are just jerks. This means we have to design not only to handle regular users, but also the jerks.

In the case of working with user reviews, I'm sure that some marketing type has demanded that users be able to not only upload plain text like, "Wow, this merchant is great, I got all my stuff in just 24 hours, I'd buy from them again!" but also to be able to use HTML like, ¡b¿¡font color="red"¿Wow!¡/font¿¡/b¿, which is full of bold and red, and if they could get away with it, dancing GIFs, because marketing people seem to get paid based on the number of incredibly stupid features they add to a project. I direct this comment not at all marketing people, just those who think that an interface with 20 buttons is far better than one with 10. I believe Dante wrote about such people and that there was a special level of hell for them. The problem before you is how to let some subset of HTML through, at least for the bold, underline, and perhaps colors, and to not allow anything else. The approach you're looking for is a white list and in pseudo-code a function to clean up a string to only allow these tags looks something like the following:

```
//
// Function: string_clean
// Input: an untreated string
// Output: A string which contains only upper and lower case letters,
//         numbers, simple punctuation (. , ! ?) and three types of HTML tags,
//         bold, italic and underline.
//
string string_clean(string dirty_string)
{

string return_string = "";

array html_white_list = ['<b>', // bold
   '<i>', // italic
   '<u>']; // underline

array punctuation_white_list = ['.', ',', '!', '?']

for (i = 0, i < len(dirty_string), i++)
{

if (isalpha(dirty_string[i])) {
return_string += dirty_string[i];
continue;
} else if (isnumber(dirty_string[i])) {
```

```
return_string += dirty_string[i];
continue;
} else {
if (dirty_string[i] is in $punctuation_white_list) {
return_string += dirty_string[i];
continue;
} else if (dirty_string[i] == '<') {
$tag = substring(dirty_string, i, i + 2);
if ($tag in $html_white_list) {
return_string += $tag;
} else {
return_string += ' ';
i += 2;
}
}
}
return_string += ' ';
}

return return_string;

}
```

The `string_clean` function has several features I'd like to point out. The first is that it is very strict, probably stricter than you'll be able to get away with when dealing with marketing, but I wish you luck. The allowed characters are all the upper- and lowercase roman alphabetic characters, all ten digits, and then four types of punctuation: periods, commas, question marks, and exclamation points. No parentheses and no braces are allowed, which protects against the case of ?{ getting through. In terms of HTML, only three tags are allowed, bold (`<b>`), italic (`<i>`), and underline (`<u>`). The function is implemented as a white list, which means that only the allowed characters are appended to the returned string. Many string cleaning routines are implemented as black lists, which is to say they list what is *not* allowed. The problem with black lists and white lists was treated in the letter from Input Invalid, so I won't go over the details again here. For those interested in efficiency note that we check for the most common case first, a letter, the next most common, a number, and then the least common cases, which are punctuation and finally the allowable tags. I picked this order so that the code would append the character and go round the loop most quickly in the most common cases, which hopefully gives us the best possible performance. You should also note that the default action is to ignore the input character and to simply append a space to the return string. We append a string so that it is possible to see where there might have been illegal text. Simply removing the offending character makes it too easy to miss where the attack may have been.

Of course, this is a simple first pass at a filtering function, and it would have to be tailored to your environment, but I hope it gives you a shove in the right direction. In order to protect against such attacks you must not only code such a function, but you and everyone on your team must use it for each and every case of user input. I cannot count the number of times that a suitable filtering function existed in a library and yet, for some perverse reason, the engineers working on the product decided to simply ignore it or to go around it because they felt they were better at treating the input themselves. I have one piece of advice for such people, don't do it. If you require some special abilities with a particular piece of input, then either extend the function or create a new one, one that can also be used consistently. It will save you a lot of time and headaches in the long run.

KV

3.11 Phishing and Infections

> *Using encryption on the Internet is the equivalent of arranging an armored car to deliver credit card information from someone living in a cardboard box to someone living on a park bench.*
>
> Gene Spafford

Phishing is perhaps the least technical of the attacks that we have to protect against using technical means, for phishing is, in reality, more like a human to human con job; rather than being an attack on code, it is an attack on the system component between the keyboard and the chair.

People have been running con jobs on each other since there were two people and one wanted to get something from the other via less than honest means. Phishing is just the ability to trick people using computer systems connected to a global network, which broadens the attack and reduces the individual risk of getting caught, since the attacker and the attacked do not have to meet in the real world.

Many attempts have been made to make phishing harder via technical means, such as validating cryptographic signatures on e-mail, having web browsers block known "bad" sites, changing the color of their address bar, displaying a lock to show a properly authenticated and encrypted session, and many other things, none of which have really made a dent in the phisher's ability to fool some of the people some of the time.

Since phishing is such a human endeavor it's actually the humans that we need to fix, although as the letter and response in this section show, there are some technical things we can do to make the phisher's life more difficult.

It might seem obvious that after over a decade of stories about phishing in the news, and for those of us in the corporate world, too many awful, animated "trainings" about phishing, that people would have become naturally more suspicious, but this has not been KV's experience. It often seems as if there are people who are naturally suspicious and those who are not. I have often told the story of my mother, who was not someone who worked with computers, and how she handled her e-mail:

    KV: I was thinking of getting you a new computer.

    MOM: I like my machine, it's fine.

    KV: Sure, but it must be full of viruses by now.

MOM: No, no problems.

KV: How can that be?

MOM: I don't browse random sites on the Internet, and when I get a mail from someone I don't know, I immediately delete it, without opening it, and then empty the trash to make sure it's gone.

KV: ...

I think back to this conversation a lot because I wish that half the people I've met in technology positions did the same and because it reminds me that part of the proper mindset to avoid phishing has nothing to do with technical know-how; it is simply that suspicious way of thinking that I was immersed in from childhood and that my entire family shares. How can one pass this sort of thinking on other than by bringing them up in a family of admitted paranoids? Perhaps we can use the immortal words of General Jack D. Ripper:

> I want to impress upon you the need for extreme watchfulness. The enemy may come individually, or he may come in strength. He may even come in the uniform of our own troops. But however he comes, we must stop him. Now, I'm going to give you three simple rules: First, trust no one, whatever his uniform or rank, unless he is known to you personally; Second, anyone or anything that approaches within 200 yards of the perimeter is to be fired upon; Third, if in doubt, shoot first then ask questions afterward. I would sooner accept a few casualties through accidents rather losing the entire base and its personnel through carelessness.

which for the case of phishing means we teach people two important things:

Trust no one, even if they're personally known to you. If you get a request to do something, verify that request via a different channel; e.g., if you get an e-mail, call them.

Shoot first and ask questions afterward. For e-mail communications this means to delete an e-mail you think to be phishing. If it's really important, and actually valid, the person will try to get ahold of you again, and at that point if you're still worried, you can contact them via an alternate means.

One topic I didn't cover in my response, which I should mention here, is the idiocy of most password recovery questions, as this comes into play once someone knows they've been phished. It boggles KV's already addled mind that in 2020 there are still systems that ask questions that are easy to glean from public records, or online searches, such as mother's maiden name and locations where you've lived. One colleague in the security world treats whatever question is asked as if it asked, "What is your philosophy of life?" and fills in something witty and memorable, only to him, into the field, so that if an attacker ever has to talk to a customer service agent to steal an account, it would be quite difficult. Of course, given the number of people who use "password" as their password, which is depressingly large, this won't help everyone, but it would be useful to those of us who actually care about our online security.

If we can somehow get these ideas to stick in people's heads, the problem of phishing will be greatly reduced, but until that day I guess we'll have to color our URLs and throw up large warnings and hope for the best.

Dear KV,

I noticed you covered cross-site scripting a few issues back, and I'm wondering if you have any advice on another web problem, phishing. I work at a large financial institution, and every time we roll out a new service check box the security teams come down on us because either the login page looks different or they claim that it's easy to phish information from our users using one of our forms. It's not like we want our users to be phished, we actually take this quite seriously, but it's also, I don't think, a technical problem; our users are just stupid and give away their information to anyone who seems willing to put up a reasonable fake of one of our pages. I mean come on, doesn't the URL give away enough information?

Phrustrated

Dear Phrustrated,

Ah, yes, your users are stupid: they just sit around waiting for someone to pop up a login screen or a page full of inputs for personal information and fill them out; they're doing it just to get you. It's very comfortable to think along these lines because it leaves you feeling superior and means you don't have to do any work to fix the problem; instead you think you should fix the users. Unfortunately, as I have learned from long experience, beating stupid people doesn't make them any smarter.

Now, I don't like users any more than you do; they're demanding and want things simple and just ruin my fun at playing with what really are "my toys." Alas, we're both paid to actually make these toys work well in the service of the users and so we have to give them some small consideration. So, what is phishing? Well, first of all it's a very annoying misspelling, far more annoying than "kode" for code. Go ahead, call me a hypocrite, I'll wait.

More to the point, phishing is the ability of an attacker to get someone to give away important or useful information. At the highest level this is one of the oldest tricks in the book and likely dates back as far as the oldest profession. It's a con job, your users are the marks, and unless the marks are paying attention, they are going to be conned out of their username and password, or Social Security number, phone number, birthday, etc. The Internet has amplified the abilities of the old-time con man (there must be also con women, but that word is not listed in my dictionary) because now there is a tremendous amount of important information stored on computers and because the Internet reaches into hundreds of millions of homes from anywhere on Earth.

Although there is no sure technical fix to the phishing problem, there are ways to evaluate possible solutions, and these ought to be kept in mind for those moments when someone in a meeting says, "Well, if we just…" I have an allergic reaction to that particular phrase because it is usually followed by a specious or poorly thought out suggestion that sounds good until you try to follow it to a logical conclusion.

Obviously, there are a lot of smart people thinking about this, but the best advice I've seen on evaluating anti-phishing technologies comes from Rusty Shackleford. Rusty's rules can best be summed up in this way:

- Anything the attacker can see, the attacker can spoof.
- Anything the user knows, the user can, and will, disclose.

    Corollary: Anything the user's browser knows, the user's browser can and will (quite willingly) disclose.

- Your solution is only as good as its first step. That is, your solution is only as good as what your users have to do when they find themselves in unfamiliar surroundings.

Let's take these one by one. "Anything the attacker can see, the attacker can spoof." It may seem simple, but many people miss this point. Often people go to a lot of trouble to put visual cues into pages so that the user "knows" they're logging into the right page. The problem is that anything you show to all your users, all the "bad guys" can see as well, and pretty easily re-create, no matter how complex they are. Eventually all that complexity just gets lost to the user anyway, so don't bother; they're not going to notice. Now, if you can come up with something that personalizes the page, perhaps an image or a picture, and not one chosen from a list of 10, or 100, then that might begin to provide some protection. Sounds are another way to personalize a page, though an annoying one for those of us who hate noise in public places like cafes or cubicles.

A harder problem to tackle is the one that "Anything the users knows, the user can, and will, disclose" with its corollary about the user's browser being tricked in place of the user themselves. Let's face it, this is the root of the problem; the users are being tricked, and if an anti-phishing system just depends on a different set of data being collected from the user, for instance the question "What is the meaning of life?" instead of "What is your Mother's maiden name?" then you're just moving the problem around, like shuffling the deck chairs on the *Titanic*. One of the goals of a good anti-phishing system should be, if possible, to do away with collecting any "personal and secret" information from the user because that information isn't really personal or secret and they'll willingly type it into a cleverly built phishing page.

Perhaps the hardest of the rules to understand is, "You're only as good as your first step." What Rusty was saying here is that no matter how good and tightly constructed the later phases of your system, the whole thing will unravel if the first step is susceptible to any of the issues pointed out in the previous two rules. Confused users are the easiest ones to phish. So, for example, if your account recovery page requires the user to type in a lot of complex "personal and confidential" information that the user knows, it is likely that the phisher is going to take advantage of this, and instead of hosting a fake login page, they'll happily host a fake account recovery page. It's just as easy to steal an account with the account recovery information as it is with the login and password.

Alright, I admit it, I was unable to solve phishing in 1,200 words or less, but I hope Rusty's advice is a help to you and makes you a little less frustrated. It's certainly allowed me to quash some of the more questionable anti-phishing ideas I've heard of, and some days, that's half the battle.

KV

3.12 UI Design

> *A common mistake that people make when trying to design something completely foolproof is to underestimate the ingenuity of complete fools.*
>
> Douglas Adams

What can a low-level systems person really know about UI design? Well, I may not know much about UIs, but I know what I like, which is to say I know a lot about what I don't like. The following letter and response aren't so much about how to design a UI but how to insulate the design of the UI from the design of the overall system. Many modern user interfaces are built on the Model-View-Controller and Model-View-Presenter paradigms, which do an adequate job of insulating how the system looks from how it does the job it needs to do. The model is meant to contain the data, and the controller the logic for manipulating that data, while the view or presenter presents that data to the user in whatever way the UI designer thinks is most appropriate. Keeping the UI designer in a box where their choices cannot negatively affect the overall logic of the system is often best; sometimes I even drill air holes into that box for them, but sometimes I do not.

The following letter and response do not discuss a formal paradigm such as MVC or MVP but do touch upon the important point, especially for those doing large systems design, of trying, as much as possible, to create a clean interface between how the system looks to the user and what it does for the user, for this type of fence makes for the best type of neighbors.

Dear KV,

I've been reading your column occasionally in Queue, and I haven't seen you address anything related to user interface design and how it can completely torque a piece of software. I happen to work as a programmer on a project for a company that sells point-of-sale software, which is a nice way of saying cash registers. The goals of the marketing people and user interface designers, and we have several for our different product lines, always seem to twist the software into directions that only make it more fragile. These people ask for features that while to a naive user might make the user interface easier to customize or use, to any of the programmers on the project it's plainly obvious that the feature in question will have a negative impact on code size, clarity, or some other nasty side effect. Several times our releases have been delayed because mid-stream we were asked for a feature that proved to have such a horrible side effect that it had to be removed right at the end. Sometimes these "features" are really just visual changes, but our system is so easy to change visually that it seems to invite the marketing and design folks to change just for the fun of it, as if the color of the buttons ought to be red one day and then blue the next. Is there any way to make these pests go away?

Torqued

Dear Torqued,

What's this that you "...only read me occasionally!?" Well, if you had been reading all the time, you'd know that I haven't actually addressed your issue, but I'll let you in on a little secret. Long before my lucky escape from the presentation layer I thought I too would like to work on user interfaces. After all, they're the first thing you see when someone uses your software, and done right they are the first thing that people praise. Very few people come up to you and say, "Hey, nice protocol, I really like the way you made use of that spare bit in the flags field." It was actually quite a big kick the first time I could say to my mom, (yes, I have a mom and was not hatched as so many people seem to claim), "Hey, see that? I did that!" and could explain in simple terms what the thing I did was doing. Two things took me away from UI work: the first was a deep interest in the lower levels of operating systems such as device drivers and networking, and the second was the same marketing and design people you mention. I remember one, who seemed to almost delight in asking for changes as if the accumulation of modifications was his personal contribution to the system we were building, which they weren't; they were just a drain on resources. Alas, this particular person wasn't fired out of a cannon, as I often asked management to do, nor even fired; he just took another job, I suspect in order to find someone else to try and torture, or because I kept slashing his tires. I still wonder if he ever figured out who it was who did that.

Now, before I go on, let me say that there are some wonderful user interface designers who really get that it takes more than a few seconds to make pervasive changes in software and who, therefore, choose their requests carefully and work with the programmers to get things into a suitable state. There are five of them, and no, I will not give out their e-mail addresses.

Over the years I have developed a few strategies for dealing with the more interfering types of designers, who aren't really designers at all; they're just quacks who can talk smoothly about the use of color and are fine for helping you decorate your house but useless at software. Oh, and none of these strategies involves violence or will get you arrested, I think, so I feel that I can share them.

One of the most important things a programmer or software architect can do in the early stages of building a system, to protect themselves from being yanked around, is to separate the way in which information is presented to the user interface from how it is stored and manipulated in the real system. I know that such a rule must seem obvious, but I have seen people do the exact opposite time and again. It is a fine idea to look at what data the system needs to store and manipulate as a set of inputs, but how it is stored and manipulated must be separate from how it is input or you are already lost. Dante seems to have left out this particular region of hell in *The Inferno*, likely because such sins were less prevalent before computers and user interfaces became ubiquitous, but it is there nonetheless. I know, I've been. I got the scars and the company T-shirt to cover them with.

OK, now that we've got the presentation and manipulation systems apart, the next thing to do is to design a presentation layer that is easy to change without breaking the rest of the system. If your designer wants to change the color, text, font, font size, and character set used throughout the user interface on a weekly, daily, or hourly basis, let them! It gives you more time to code real features if they're off playing with Pantone color wheels, fonts, and button borders. Make sure that changing the user interface doesn't require any knowledge of building the software or using any complex tools like text editors. The last thing you want is to babysit, as I have, the designer as they complain that every time they build the code, it breaks. It should be completely unnecessary to compile anything to change the user interface.

Making sure that the user interface can be simulated in the absence of a real system is often a big win as well. You'll be much happier if the designers can play with the UI while you're building the rest of the system instead of dealing with the constant maintenance of the unfinished system because the designers need some feature *right now* that you know you don't need until much later.

With all this decoupling going on, the biggest internal danger to watch out for in your system is layer proliferation. A favorite quote of mine, attributed to Van Jacobsen, a networking researcher, is, "Layers are a good way to think about network protocols, but a poor way to implement them." The same is true for most software. How many layers do you need? Enough to get the job done and not so many that they impact performance. Don't like squishy answers like that? Tough, suck it up. It's your system, and you'll have to deal with this on your own, but if you don't think about it at all, you will really hate yourself later. Either you will wind up with too few layers, and therefore nasty tendrils of code violating the layers and causing problems, or you'll wind up on the other end of the spectrum where you might as well move the data on paper because it will be faster than waiting for it to crawl its way up from the murky depths of secondary storage only to see the light of day minutes later. Neither is a system you want to work on, trust me.

So, there you have it, KV's quick list of how to avoid getting your chain yanked by the color selection brigade: separate the API from the UI, build a presentation layer that can be changed without any need to be a programmer, simulate the back end as soon as you can, and make sure you have the right number of layers.

KV

3.13 Secure Logging

> *The most secure code in the world is code which is never written.*
>
> Colin Percival

Logging systems are often the Achilles heel of security. A very common interaction when reviewing a system can be paraphrased in the following way:

KV: What kind of data does the system store?

SUSPECT: Some personal information.

KV: [*Raises eyebrows, lowers voice, speaks more slowly*] Such as...?

SUSPECT: Oh, you know, name, address, phone number, e-mail...

KV: And where is this stored?

SUSPECT: We store it in our database.

KV: [*Lowers voice even further, to a dangerous whisper*] Encrypted?

SUSPECT: [*Jumps as if bitten*] Of course!

KV: [*Voice returning to normal*] Good, good. What about payment information?

SUSPECT: We store that in a separate database, also encrypted.

KV: Great! Now, does the system log transactions for debugging and tracing down problems?

SUSPECT: Of course!

KV: Please tell me what data is logged for each transaction.

SUSPECT: We usually run with a high logging level so it's easy to track down problems. At that level we log all the information for each transaction.

KV: Including the personal information?

SUSPECT: [*Starts to look worried.*] Yes...

KV: [*Asks in an off-hand manner*] And the credit card info?

SUSPECT: [*Looks at shoes*] Yes...

KV: [*Takes off glasses, rubs bald head*] In plain text.

SUSPECT: Uh, yes.

This is not an exact transcript, because an exact transcript would have a lot more colorful metaphors right at the end.

The point of the dialogue of course is the point we often come to when people think they are securing a system; they only lock the front door, but they leave the windows, cellar, and back doors wide open. Building a secure system doesn't just mean following a set of guidelines or a run book; it means thinking about all the places where data is accessed and how that access is controlled. Logging systems are just a very common form of leaked data, but there are plenty more, for instance debugging interfaces, which are another common way to access a system that is frequently left open in shipping systems.

With the advent of cheap and easy to use cryptographic filesystems, much of the storage of such log data should now be less of a problem, but it is still rare to see these in practice.

Building a secure logging system is only one part of building a secure system, but as logging systems are so common, let's start there.

Dear KV,

I've been stuck with writing the logging system for a new payment processing system at work. As you might imagine, this requires logging a lot of data because we have to be able to reconcile the data in our logs with our customers and other users, such as credit card companies, at the end of each billing cycle, and also if there is any argument over the bill itself. I've been given the job for two reasons, which are that I'm the newest person in the group and because no one thinks writing yet another logging system is very interesting. I've also not gotten a lot of help from the other people on the team who claim to have, "written far more logging systems in their time than they want to think about." Any advice on doing up a proper logging system?

Logged Out

Dear Logged Out,

If so many of your teammates have written logging systems before, how come you're not using those? Perhaps your teammates are lying to you and have never written a single line of a logging system, or perhaps, and I suspect this is probably more likely, they tried and their systems sucked rocks. Or perhaps I'm just cynical.

It turns out that writing a good logging system, like writing any good piece of software, is both difficult and rare. Many of your decisions are going to depend on the requirements put on the data you're logging, and since you're logging financial transactions, you have a lot of requirements, some of which must include the ability to keep the data private, to audit the log for errors, and to be able to verify that the data contained in the log has not been tampered with.

Data privacy is now a big deal in our industry. It's too bad that it wasn't a big enough deal to the companies made famous in the last few years for breaching private data, such as Choice Point, Bank of America, Wells Fargo, and Ernst and Young, but they're all smarting for it now. Personal data breaches are now such a big problem that several governments have enacted strong legislation to punish those offenders, and I think you'd like to avoid such punishment; I know I would.

The best way to keep data private is not to store it at all. Storing data makes it possible to breach it, which seems obvious, but then again every time I think something is obvious I wind up reading a news item that tells me, no, not obvious enough. Only keep whatever data you need to back up whatever claim you need to make, and don't keep data for too long. Most financial institutions have limits on how long they'll keep data; follow the relevant ones for your product to the letter, and don't keep anything a second longer than you need it.

Once you've winnowed down the list of things you actually need to keep in the log, decide which ones can be blinded, which ones must be encrypted, and which can be left in the open. Blinding data means that it is destroyed, but in a way that makes it unique. A hash function is a great way to do this. Given any input, a good hash function produces unique, seemingly random, output. Consider the following example using the md5 program on my Mac:

```
? md5 -s "1234 5678 9012 3456"
MD5 ("1234 5678 9012 3456") = d135e2aaf43ba5f98c2378236b8d01d8

? md5 -s "1234 5678 9012 3457"
MD5 ("1234 5678 9012 3457") = 0c617735776f122a95e88b49f170f5bf
```

Given two strings, which look like fake credit card numbers, where only one digit is different in one position, the md5 program produces what look like two different random numbers. If you can find a pattern in these, please contact your local member of MI6 or equivalent as they have a job for you in their signals department.

Not only are these two numbers seemingly random, but they are also unique, which means they make a fine primary key for using in your data logging. Each log entry with these numbers uniquely identifies the credit card, but someone reading the log cannot figure out the original credit card number from the hash. Blinding can be used on all kinds of data, but it's definitely good to use it on things that if they were stolen or compromised could be used by others.

If there is data that you absolutely must be able to use again in its original form, that is, it cannot be blinded, then it's time to start encrypting, at least if that data is valuable in any way. I am amazed at the number of people who go to great lengths to properly encrypt data in their databases and live systems but then just chuck it all, unceremoniously, in plain form, into the logs. I guess I should stop being amazed, but it's preferable to banging my head on the desk, wall, floor, or engineer in question.

What kind of data might need to be kept secret in your logs? An exhaustive list isn't possible, but certainly personal details like the person's full name, address, phone number, mobile number, and e-mail address are a good start. While you're at it, the amounts paid, locations of payment, and other payment specifics should also be kept secret as they make your logs a juicy target for people trying to dig up financial data on your company. You might be asking, "Well, what's left?" and I have to say that in a financial system, probably not a lot, but I'm sure there is data around for debugging purposes that might be OK to go into the log in plain form. For instance, the time the entry was made is probably not going to be secret.

Now that you've eliminated all extraneous data, blinded what you could, and likely encrypted most of the rest, you have to make sure that the log itself is secure from tampering. You will need to do two things in order to prevent tampering, sign the entries and sign the log, each with a different key. The entries are signed to ensure their validity, and the whole log is signed to make sure that no one has added or removed entries by hand. The reason two different keys are used is that two different people should have those keys, thereby requiring collusion to violate the security of the system. It's also a good idea to change your keys regularly so that if a key is stolen, the amount of data that is exposed is minimized.

There are many other things to touch upon with a logging system, such as where the data is stored, how it may or may not be moved across a network, when the logs need rotating, and how to write tools to analyze and read the log, but what I've given here are the basics of making a logging system that I hope doesn't suck and doesn't make it trivial to violate the privacy of your users and land your company on the front page of the news. Oh, and one last piece of advice, don't leave the logs on a laptop in your car. Obvious? Sure, it's obvious.

KV

3.14 Java

> *If Java had true garbage collection, most programs would delete themselves upon execution.*
>
> Robert Sewell

I have been lucky enough in the years between receiving this letter and writing the response to, mostly, have avoided the scourge that is Java. You will find that in these pages I often extol the idea that every language has a domain for which it is best suited, but that liberal ideology doesn't mean that I don't have opinions on languages when I see the damage they have wrought. It is nearly laughable to me that the original impetus for Java was embedded systems, where it was thought to be "Write once, run everywhere," but as someone who works with embedded systems daily, I can say that I very rarely come across Java. Having escaped from its creators' original intentions, Java literally went everywhere…servers, browsers, and, currently, applications running on the Android system in phones. The use of Java in Android now means that there is a generation of koders who, in order to write their applications, have been forced to use this language, one that often results in people writing code as described in the following letter and response. The only thing worse than the fact that Java is part of Android is that during the late 1990s and early 2000s many undergraduates were taught Java as their introduction to programming, a mistake of truly mind-bending proportions, for while Java contained all that was supposed to be good in software engineering at that time (classes, objects, methods, etc., etc.), it did not have a way to gently introduce people to these concepts. The moment you touched anything in a Java system you got all the concepts rained down upon you like hell fire. As teaching languages go, Java is the last one I would foist upon new students, but many departments felt the industry wanted programmers trained in the latest language, rather than having them learn to program intelligently so that they could program in any language.

As you can see from my comments, my love of Java has only grown since KV was first asked, and replied, about this grand creation that hopefully someday will go the way of the dodo. People remember that the dodo went extinct, but they don't always remember that that extinction was brought about by conscious action by the hand of man.

Dear KV,

You mention Java in passing.[2] I've done a day's intro and got a book on it but never had to produce any serious code in it.

In an admin role, I've been up close and personal with a number of Java server projects, and they seem to share a number of problems:

- Performance. About 10 times slower than C.
- Complexity. Seems very large and obtuse.
- Slow to code. Always running behind.

My observation is that Java *requires* a GUI/IDE to code effectively—that ordinary mortals can't keep all the classes and their APIs in their close cognitive span (at their fingertips). Compare this to Perl, which doesn't have a decent IDE available. I'd suggest it's not because people of talent or interest aren't available, but because there is no driving need for it. It's simple to get into and efficient to code with a text editor. And it's *not* a language that's suitable for large projects. Maybe Perl 6 will be. ;-)

Is there *any* data that says that Java projects are any more or less successful than older languages? It's got heavy commercial support and lots of press and has noble aims of helping programmers and reducing certain classes of faults... But as *professional* programmers, we use sharp tools, and they *are* dangerous for exactly the reasons they are useful. Trying to protect everyone from "level 1" programmer errors seems very limited to me.

I keep seeing projects to replace "legacy" apps start amid fanfare and hoopla—and significant budgets—using "the most modern techniques" that end up being cancelled or only partially implemented.

Am I missing something??

Run Down With Java

Dear Run Down,

Having taken a course on Java and read a book on it, you're actually ahead of old KV on the Java wave. I'm still hacking C, Python, and bits of PHP for the most part. Given your comments, perhaps I'm lucky, but somehow I doubt that; I'm rarely lucky.

In a way I could almost reprint your letter without comment, but I think there are larger issues that you raise, and I really can't let these things go without commenting, or, perhaps, screaming and tearing my hair out. It turns out that shaving my head has helped with all those bald patches I got from tearing my hair out.

---

2. George V. Neville-Neil: Gettin' Your Kode On, in: ACM Queue, Feb. 2006.

As a reader of KV you've probably already realized that I rarely bash languages or make comparisons between them, and I'm going to stick to my guns on that, even in this response. I don't believe the majority of the problems you're seeing come from Java itself, but how it is used, and also the way in which the software industry works at this point in time.

The closest I've come to Java was to work on a project to build some lower-level code in C that would be managed by a Java application. There were two teams, one that wrote the systems in C, which could operate independently of the Java management application, and one that wrote in Java. Now, you would expect that the Java team and the C team would have met on a regular basis and that they would have exchanged data and design documents so that the most effective set of APIs could be built in order to efficiently manage the lower-level code. Well, you would be wrong. The teams worked nearly independently, and most of the interactions were disastrous. There were many reasons for this, some of which were traditional management problems, but the real reason for this "failure to communicate" was that the two teams were on two different worlds, and no one wanted to string a phone line between them.

The Java team members were all into abstraction; their APIs were beautiful creations of sugar and syntax that scintillated in the sunshine moving everyone to gaze in wonder. The problem was that they didn't understand the underlying code they were interacting with, other than to know what the data types and structure layouts were. They did not have a deep appreciation of what their so-called management application, was supposed to manage. They made grand assumptions, often wrong, and when they ran their code, it was slow, buggy, and crashed a lot.

The C team members weren't all perfect either. There was a certain level of arrogance, shocking I know, toward the Java team, and although information wasn't hidden, it was certainly the case that if a C engineer thought one of the Java engineers didn't "get it," they would just throw up their hands and walk away. The C team did produce working code that shipped, worked well, and didn't fall over. The problem was that the goal of the company was to build an integrated set of products that could be managed by a single application. Although the C team won the battle, the company lost the war.

Someone looking at the code as it was delivered might have thought, "Well, the Java programmers just weren't up to the task, next time hire better programmers, or get better tools or..." The fact is that that's not the real problem here. The problem wasn't Java; it was the fact that the people building the system could produce a lot of lines of code but didn't understand what they were building.

I have seen this problem many times. It often seems that projects are planned like some line from an old Judy Garland/Mickey Rooney musical. One character says to the other, "Hey, kids, let's put on a show!" It always works in the movies, but as a project plan it rarely leads to people living happily ever after.

In order to build something complex you have to understand what you're building. The legacy applications you mention are another great example. Ever seen a company convert a legacy app? I hope not, it's not very fun. The way legacy conversion goes is that you have a program that works. It does something. You may have the source code or you may not. No laughing now, I've seen this. When the legacy program runs, it does what it should, most of the time. Next the team comes in and tries to dissect what the program does and then reproduce it, with bug for bug compatibility, and finds that their modern techniques don't reproduce the same bugs in the right way and so they get to a point where the program sort of works, or sort of doesn't, and then usually give up and re-implement whatever it was, from scratch.

One of the reasons such travesties can continue to occur is that unlike in any engineering discipline in the real world, like aeronautics, or civil engineering, failure just means a loss of money. Now, when I say "just," that can be a big just. The overhaul of the U.S. Internal Revenue Service computer systems cost millions in overruns as did the system developed for the Department of Motor Vehicles in California. There is a laundry list of such failed projects to choose from. These may make headlines for a while, but they're not quite on the level of a bridge failing, like the Tacoma Narrows, or the Space Shuttle exploding, twice. People generally remember where they were when the space shuttle Challenger blew up, but they don't remember where they were when they heard about an IRS computer cost overrun.

With more and more computers and software being put into mission-critical systems, perhaps this attitude will change with time. Unfortunately, it's going to require a few more spectacular failures, likely with a human instead of monetary cost attached, before people put more time into planning what they do and figuring out what their code is actually meant to be doing. Once we do that, then the fact that we're using Java or Perl or the language du jour will have a lot less effect and probably be discussed a lot less as well.

KV

3.15 Secure P2P

> *When you pirate MP3s, you're downloading COMMUNISM.*
>
> —Anon

The following piece truly takes me down memory lane, to back when peer-to-peer (P2P) systems were looked upon in the same way block chain is today. Those were the days of Napster and the exchange of material, copyrighted and otherwise, from one computer to another without much reliance on central servers. Today there are plenty of systems that operate in a peer-to-peer fashion, and their legality and legitimacy is unquestioned. The fact is that peer-to-peer systems continue to be used to exchange data that annoys lawyers with mouse ear hats, but those uses have now been overtaken by other applications that are considered more respectable, or at least are backed by VC money.

As is often said by koders when we get, inevitably, stuck talking about topics that stray into law, IANAL (I am Not A Lawyer), and the response here quickly morphs into a discussion of secure file sharing, which might even be the start of some not unreasonable ideas on securing distributed systems, since that's all, from a technical standpoint, a P2P system is. They are systems that do not, as much as possible, depend on a centralized infrastructure to achieve their goal but instead spread the work among a set of clients, where work is done by each according to its ability, and service is provided to each according to its needs.

Dear KV,

I've just started on a project working with P2P software, and I have a few questions. Now, I know what you're thinking, and no, this isn't some copyright-violating piece of cowboy code, it's a respectable corporate application for people to use to exchange data like documents, drawings, and work-related information.

My biggest issue with this project is security, like accidentally exposing our users' data or leaving them open to viruses. There must be more things to worry about, but those are the top two.

So, I want to ask, "What would KV do?"

Unclear Peer

Dear UP,

KV would run, not walk, to the nearest bar and find a lawyer. You can always find lawyers in bars, or at least I do; they're the only ones drinking faster than me.

OK, so let's assume your company has lawyers to protect them from the usual charges of providing a system whereby people can exchange material that perhaps certain other people, who also have lawyers, consider it wrong to exchange. What else is there to worry about? Plenty.

At the crux of all file sharing systems, whether they are peer to peer, client/server, or what have you, is what type of publish/subscribe paradigm they follow. The publish/subscribe model used defines how users share data.

The models follow a spectrum from low risk to high risk An example of a high-risk model is one in which the application attempts to share as much data as possible, such as sharing all data on your disk with everyone as the basic default setting. Laugh if you like, but you'll cry when you find out that lots of companies have built just such systems, or systems that are close to being as permissive as that. Here are some suggestions for building a low-risk peer-to-peer file sharing system.

First of all, the default mode of all such software should be to deny access. Immediately after installing the software, no new files should be available to anyone. There are several cases of software that did not obey this simple rule, and so when a nefarious person wanted to steal data from someone, they would trick them into downloading and installing the file sharing software; this is often referred to as a "drive by install." The attacker would then have free access to a person's computer or at least to their MyDocuments or similar folder.

Second, the person sharing the files, that is, the sharer, should have the most control over the data. The person connecting to the sharer's computer should only be able to see and copy the files that the sharer wishes them to see and copy. In a reasonably low-risk system, the sharing of data would have a timeout such that unless the requester got the data by some time, say, 24 hours, the data would no longer be available. Such timeouts can be implemented by having the sharer's computer generate a one-time-use token containing a timeout that the requester's computer must present to get a particular file.

Third, the system should be slow to open up access. Although we don't want the user to have to say "OK" to everything, because eventually they will just click OK without thinking, you do want a system that requires user intervention to give more access.

Fourth, files should not be stored in a known default location or one that is easily guessable. Sharing a well-known folder, like MyDocuments, has gotten plenty of people into trouble. The best way to store downloaded files and shared files is to have the file sharing application create and track randomly named folders beneath a well-known location in the filesystem. Choosing a reasonably sized random string of letters and digits as a directory name is a good practice. Why should you bother with creating hard-to-figure-out folder names? Because it makes it harder for virus and malware writers to know where to go to steal important information. If the filename and path are well known, it's far easier for someone to steal data from the machine.

Fifth, and last for this particular letter, the sharing should be one to one, and not one to many. There are many systems that share data one to many, including most file swapping applications, such that anyone who can find your machine can get at the data you are willing to share. Global sharing should be the last option a user has, and not the first. The first option should be to a single person, the second to a group of people, and the last, global.

You may note that a lot of this advice is in direct conflict with some of the more famous file sharing, peer-to-peer systems that have been created in the last few years. The reason for that is that I have been trying to show you a system that allows for data protection while data is being shared. If you want to create an application that is as open, and also as dangerous as Napster, or its errant children, were and are, then that's a different story, but it sounded from your letter that that was not what you wanted.

Other things you will have to worry about include the security of the application itself. A program that is designed to take files from other computers is a perfect vector for attacks by virus writers. It would be unwise, well, actually, it would be incredibly stupid, to write a program such that it executes or displays files immediately after transfer without asking the user first. I have to admit that answering "Yes" to "Would you like to run this .exe file?" on Windows is about the same as saying, "Would you like me to pull the trigger?" in a game of Russian Roulette.

Another open research area, er, I mean, big headache, which I'll not get into here, is the authentication system itself. Outside of all the other advice I just gave this problem is itself quite thorny. How do I know that you are you? How do you know that I am me? Perhaps I'm the Walrus…

KV

4

Distributed systems are hard.

Jonathan Anderson

Machine to Machine

Of all the technical problems I have come across while writing Kode Vicious, none has intrigued me more than distributed systems, probably because they are fantastically hard to control and also to get correct. Decades of research has produced some usable results, Lamport's algorithm, PAXOS, and a few others, but in general this area of computing remains a dark art to many, and often it's the simplest concepts that trip people up.

Most koders have the false belief that they'll never have to design a network protocol and that this is an esoteric pursuit of a select few who work on standards committees. The fact is that anyone who is writing code that uses the network is, in effect, defining a network protocol. The process of designing a network protocol could be a book on its own. The best books about networking, though, including those by the late Richard Stevens, talk more about a protocol after it was written than how it was written in the first place. Currently the best way to learn about how network protocols are written would be to follow some of the IETF drafts, attend IETF meetings, and see, as they say, how the sausage is made. Unfortunately, that's time-consuming, and if you want to attend the meetings in person, an expensive proposition. Most of the courses you'll find about networking are, again, describing currently existing network protocols and systems, rather than how they came to be. If you want to learn how to write a protocol of your own, even if it's "just a prototype that I'm playing with and will never be used in production," then you're pretty much on your own.

The letters in this chapter span several practical topics in machine-to-machine communication and are hopefully some useful guideposts to people developing on, deploying, or just dealing with networked and distributed systems.

4.1 Stepping on Toes

> *She doesn't think she waltzes
> but would rather like to try.*
>
> from *The Mikado*

All resources are limited, including those virtual resources we create in software and systems. For networked systems one of the most precious resources is the number of ports on a system with which a program can communicate. The number of ports is limited by protocol design; the protocol designer decides how many bits are used to describe a port, and once this value is fixed, it cannot be changed without updating the protocol itself. A good example is the Transmission Control Protocol (TCP) in which the port has 16 bits for 65,356 ports, and seems unlikely to be increased any time soon as this size remained unchanged in the more recent IP version 6 protocol rework. To a human, 65,536 of anything seems pretty large, but to a computer, or to a koder, we know that in modern systems this is tiny, and since, in networking, we often have to reserve these numbers to facilitate communication, the number has been quickly eaten up as people design, promulgate, but often fail to keep these rendezvous points up to date. One good way to break networked systems is to overallocate ports, as we see in this next letter.

Dear KV,

I've been working on a personal project that involves creating a new network protocol. Out of curiosity, I tried to find out what would be involved in getting an official protocol number assigned for my project and discovered that it could take a year and could mean a lot of back and forth with the powers that be at the IETF (Internet Engineering Task Force). I knew this wouldn't be as simple as clicking something on a web page, but a year seems excessive, and really it's not a major part of the work, so it seems like this would mainly be a distraction. For now, I just took a random protocol number that I know doesn't conflict with anything on my network—such as UDP or TCP—and things seem to work fine. I guess my real question is why would anyone bother to go to the IETF to ask for this unless they were a company that could waste someone's time on an e-mail campaign to get a properly assigned number?

Waiting

Dear Waiting,

Let me begin by complimenting you on the fact that you actually went so far as to find out how one might do this correctly. (I am sure that many readers have just spit coffee on their screens, because none of them can remember the last time I complimented a writer to this column.) I compliment you because just recently I came across someone who knew the right thing to do and then did exactly the opposite.

Because of some of the assumptions present in the original design of the Internet, some parts of the IPv4 packet header are far more precious than others, and while the limitations of the 32-bit network address get the largest amount of attention, the 8-bit protocol field is equally as important, if not more so. With an 8-bit field, we can layer only 255 possible protocols on top of IPv4, which may seem like a lot, and since most people assume that all IP packets carry only TCP, protocol 6, there is plenty of space. It turns out that more than half of the numbers have been used for one protocol or another, leaving only 109 for use by authors of new protocols. Another problem is that IPv6, the nominal savior of the Internet, with its wider network addresses, still uses an 8-bit protocol field, so we're not getting any more space any time soon.

The protocol field can be seen as a commons for the Internet, so let me tell you a tragedy, and one that didn't have to happen. It is a story of hubris and zealotry and, unsurprisingly, involves the collision between corporations and open source.

Sometime in the late 1990s, a group of companies got together and proposed a protocol that would be standardized within the aegis of the IETF. It's not particularly important what the protocol does, but it's called VRRP (Virtual Router Redundancy Protocol) and exists so that two or more routers can act as peers in a fail-over scenario. If one router fails, another router discovers this via a means described in the protocol and takes over for the failing router. After the standard was published, two companies—Cisco and

IBM—both claimed to have patents to some of what the protocol did. Cisco released its claimed intellectual property under a RAND (reasonable and nondiscriminatory) license. In nonlegal terms this means that people could implement VRRP, and Cisco would not chase them down with expensive claims. RAND licenses are often used in software standardization processes.

Unfortunately, there is a segment of the open source community that is incapable of playing well with others, when those others don't play the way they want them to. For those who have not had to deal with these people, it's a bit like talking to a four-year-old. When you explain checkers to your niece, she might decide that she doesn't like your rules and follows her own rules. You humor her, she's being creative, and this is amusing in a four-year-old. If you were playing chess with a colleague who suddenly told you that the king could move one, two, or three places in one go, you would be pissed off, because this person would obviously be screwing with you, or insane.

Have I lost my mind?! What does this have to do with VRRP or network protocols?

The OpenBSD team, led as always by their Glorious Leader (their words, not mine), decided that a RAND license just wasn't free enough for them. They wrote their own protocol, which was completely incompatible with VRRP. Well, you say, that's not so bad; that's competition, and we all know that competition is good and brings better products, and it's the glorious triumph of Capitalism. But there is one last little nit to this story. The new protocol, dubbed CARP (Common Address Redundancy Protocol), uses the exact same IP number as VRRP (112). Most people, and KV includes himself in this group, think this was a jerk move. "Why would they do this?" I hear you cry. Well, it turns out that they believe themselves to be in a war with the enemies of open source, as well as with those opposed to motherhood and apple pie. Stomping on the same protocol number was, in their minds, a strike against their enemies and all for the good. Of course, it makes operating devices with both protocols in the same network difficult, and it makes debugging the software that implements the protocol nearly impossible.

In the end the same thing is going to happen as happens when your four-year-old niece up-ends the checkers game in frustration. She runs away crying, and you're left to pick up the pieces. A few of us now have to take this protocol and actually get it a proper protocol number and then deal with the fact that legacy devices are still using the old, incompatible protocol.

Now I think you see why I wanted to compliment you. Doing the right thing in the commons is good for all of us. Thanks for not being a jerk.

KV

4.2 Paucity of Ports

Meet me under the clock.

Instructions for meeting at Grand
Central Station, New York City

Ports are the rendezvous points of all networked systems, and as we've seen in Section 4.1, they are a limited resource. As a limited resource, they require us to develop a few tricks to conserve them in our applications, and it is to one of these tricks we now turn.

Dear KV,

I've been debugging a network problem in what should be a simple piece of network code. We have a small server process that listens for commands from all the other systems in our data center and then farms the commands out to other servers to be run. For each command issued, the client sets up a new TCP connection, sends the command, and then closes the connection after our server acknowledges the command.

In our original data center the controller had no problems, but then we moved to a larger data center with more client machines. Now we frequently cannot make connections when trying to execute commands, slowing down the overall system. It's such a simple design that I have no clue what could be going wrong. The controller itself is only one page of code! I suspect the network gear in the new data center is to blame and that it cannot handle the load of incoming connections.

Connection Denied

Dear Denied,

Only one page of code and yet you can't find the bug? Perhaps the code has no bugs, and it is the switches. Certainly this is the usual blame game for developers: "My code worked before! You changed something!" Blame the change.

The problem is that you have to blame the correct change. Nothing is worse than venting your spleen at someone and then being proved wrong and having to admit that you were wrong. I hope you didn't vent your spleen at the network admins, because in this case you are definitely wrong.

Now I'm not going to attack your code; it is very likely as correct as you claim it to be. The blame here lies with a failure to understand TCP and how to make efficient use of it.

When your client makes a connection to the server, it uses four pieces of information to identify the connection uniquely among all the others that take place between these two hosts. These pieces of information are the source and destination IP addresses and the source and destination ports. Now, I am sure your code picks a single destination port to send to; for example, the web uses port 80, and e-mail uses port 25. Whatever you picked, this was probably not a problem.

What do you choose for the remaining bit, the source port? Most code, and certainly code that is simple and takes up only a single page, lets the operating system decide. When you let the operating system decide the source port, it picks what's known as an ephemeral port. Note that ephemeral is defined in the dictionary as meaning "lasting for a very short time," but of course short is in the eye of the beholder, or in this case, in the timer code for TCP. Whenever TCP closes a connection, the connection state remains active in the operating system in what is known as the `TIME_WAIT` state. The `TIME_WAIT` state lasts for at least two times the maximum segment length of the connection, but in a data center you're never going to have this state last for less than 60 seconds, which is the minimum value for 2MSL (twice the maximum segment lifetime) defined by the system.

OK, now you know that short in your case means one minute, so that's how long a closed socket will stick around. While that closed socket is around, the ephemeral port number that the operating system picked cannot be reused. That should be fine, you might think, since there must be a lot of port numbers available. Unfortunately, TCP was defined at a time when the idea of having millions of connections between machines wasn't really practical, or even possible, so the number of possible ports may surprise you: it's 65,536. Ephemeral ports don't get even that full set, but a subset, and on most modern operating systems the range is about 16,384.

That limit means that you can open only 16,384 connections with ephemeral ports unless you modify some of the tuning variables in your operating system. Even if you do increase the range, say to 32,768, it is quite easy for modern computers to run through that number of connections in a minute, and remember, any connection you have closed will hang around for a minute after it is closed. The reason that this bug has not shown up until now is that you have scaled your systems to the point where they are, collectively, able to use up their ephemeral ports in one minute.

While you might now have many wild ideas about extending the TCP or changing the minimum value of 2MSL, I can tell you that you should forget them all. TCP was designed to maintain long flows in the Internet, and not to be some sort of short message service. The overhead of setting up and tearing down a TCP connection should be a clear indication that using it efficiently requires putting more than a few bytes over it at a time, a lesson that the early designers of HTTP did not learn, but that's a story for another time.

The correct way to use TCP in your situation is to have a local process on each client that maintains a persistent connection to the central server. Without the constant opening and closing of sockets, you will not cycle through and use up your precious ephemeral ports.

KV

4.3 Protocol Design

> *An implementation should be conservative in its sending behavior, and liberal in its receiving behavior.*
>
> J. Postel

What are some of the things to know before creating a network protocol? Perhaps the person who wrote most eloquently about this was Jon Postel, who helped design and build the modern Internet, and who is responsible for the guidance known as Postel's law, reproduced in the epigraph that heads this section, which has been turned on its head due to security concerns, but for a research network it was the exact right advice at the time.[1] Jon was, for many years, the editor of the Requests for Comments (RFCs), which are the documents that define the generic Internet protocols, and not just TCP and IP, but hundreds of other related protocols and advice on how the protocols should be used.

Some of the advice in the following response merits more discussion than KV was able to put into the original, in particular the advice on the grouping and alignment of fields. On some CPU architectures, notably Intel's, the only downside to not aligning a field to a natural word boundary (32 bits on a 32-bit processor, 64 bits on a 64-bit processor, etc.) is a decrease in performance as nonaligned reads and writes require more instruction cycles than those that are aligned. For those who work on non-Intel architectures the penalty can actually be an error that causes the system to terminate, full stop. In the networking world the classic example is very low level, residing in Ethernet. An Ethernet header has a 6-byte source and 6-byte destination field followed by a 2-byte protocol field. On packet reception the low-level software will want to decapsulate the packet following the Ethernet header and pass it up the stack to higher layers for further processing. If the Ethernet header is naturally aligned in memory, then the following header is not; it is 14 bytes after the initial alignment, and 2 bytes off of the nearest alignment point. Reading the next packet header will cause many chip architectures to fault because they do not allow unaligned reads. While this seems like it ought to be a minor point, it has bitten nearly every koder who has had to work on embedded systems for the last 40 years. The problem can and has been worked around many times before, via offsetting of memory buffers and other jiggery pokery, but it's a story to bear in mind of the effect

---

1. Eric Allman: The Robustness Principle Reconsidered, in: Commun. ACM 54.8 (Aug. 2011), 40–45, URL: `https://doi.org/10.1145/1978542.1978557`

that field offsets in protocols can have on software. The less drastic but more general point is that unaligned reads and writes cost system performance, which is nothing to sneeze at either. The problem will only be compounded as higher-end systems have all moved to 64-bit architectures, meaning that alignment is now further off and that grouping many subfields together into a single 64-bit word has significant performance advantages.

A single letter and response is far too short to do this full topic justice, but you have to start somewhere, so we'll start here.

Dear KV,

I got into a funny argument at work on a distributed systems project. One of the people on my project is insisting that all the data we send in our new, lightweight, update system, be encoded as type/length/value instead of just type/value. Since we have a type, I don't see why length matters. We can always infer one from the other.

The protocol needs to be as light as possible because it may be transported over SMS to cellphones, as well as going over a more robust network such as the Internet.

Do you think that the length must really be included?

Wondering about the Value of Length

Dear Wondering,

It's funny how so many funny arguments have unfunny consequences. Actually, the funniest thing is that you refer to the Internet as *robust*, but I think I'll just leave that aside for now.

Network protocols and protocol design are near and dear to KV's heart, and, yes, KV does have a heart. Instead of talking only about the importance of length, I'm going to broaden the scope of my answer and provide you with KV's top five network protocol design recommendations.

Always encode a length. Since you asked about type/value vs. type/length/value encoding, I'll start there. Leaving out the length in a packet or other type of encoding is a recipe for buffer overflow disaster. If the program reading the packet or data from the wire can't figure out, in an easy way, how much data to read, this will invariably lead to someone writing code wherein he or she tries to guess just how much to read—and having to guess is bad, very bad. A poorly chosen guess usually leads to a buffer overflow, and since we're discussing a network protocol, that buffer overflow can be remotely exploited. Congratulations, your new protocol is now the Internet's new whipping boy, as script kiddy after script kiddy is able to write trivial security exploits that destroy what you built.

Protocols need version numbers. People who think, "Oh this is a one-off," need to be offed, or perhaps politely removed from the project. There is rarely any such thing as a one-off in software. You need to design and build protocols, just like software, so that they can be updated. Without a version number encoded somewhere in the protocol, it is impossible to upgrade it without guessing—and guessing, again, is bad.

Align your fields. Back in the old days when, as an older friend of mine says, "Dinosaurs roamed the earth," it was very important to squeeze every last bit that you possibly could into a packet; therefore, protocols were often designed without data alignment in mind. A program receiving a packet would have to read an entire 8-, 16-, or 32-bit entity and then play bit-twiddling games with the retrieved value to get at the one or two bits it really cared about. Although the current trend in the opposite direction—wherein people design protocols that seem intentionally to waste bandwidth—is a problem, a balance needs to be struck, and that balance ought to be struck in favor of byte-aligned fields. The alignment issue most often comes up when protocol designers decide to add flags to their packets, so that brings us to ...

Group your flags. Almost all network protocols that I know have some number of flags (i.e., single bits of information) that they need to communicate. Group the flags together so that they can more easily be byte-aligned (see item 3).

Leave room to grow. Designing a protocol to be used once is a recipe for more work—work that you wouldn't have to do if you left just a little bit of room for extensibility when you came up with the protocol in the first place. A few extra flag bits and the ability to have packet extensions at a later time are two things every protocol designer should be thinking about before releasing version 1.0. This last point is not an excuse to waste space; it is a plea to think about what happens in the future, which may be only as far away as next week.

Obviously the list could be longer, and really could be the subject of an entire book, but for now those are my top five. If you can get those right, then perhaps your protocol will become wildly popular and everyone will ask you to "do it again!" Won't that be fun?

KV

4.4 Which Came First?

> *I'm not as think as you stoned I am.*
>
> Old college adage

Once you spend enough time looking at network protocols, you realize some of them were probably designed by those famous network architects, Abbott and Costello. Abbott and Costello had an intimate understanding of network protocols, as we can see from their early design discussion reproduced here:

> COSTELLO: Look, Abbott, we have to get these packets going between these hosts, but it's hard to tell them apart. Do you think we should give them names or something?
>
> ABBOTT: They already have names.
>
> COSTELLO: They have names? Great! Tell them to me.
>
> ABBOTT: Who comes first, What comes second, and I don't know comes third.
>
> COSTELLO: That's what I want to find out.
>
> ABBOTT: I said, Who comes first, What comes second, and I don't know comes third.
>
> COSTELLO: Are you the protocol designer?
>
> ABBOTT: Yes, I am.
>
> COSTELLO: And you don't know the order of the packets?
>
> ABBOTT: I do know, Who comes first, What comes second, and I don't know comes third.
>
> COSTELLO: So tell me who comes first.
>
> ABBOTT: Yes.

While we don't give network packets names, we do give them sequence numbers, that is, if we want to have any chance of putting them all back in the correct order when we receive them. Leaving out sequence numbers, at any layer, is a big mistake, one that is covered in this next letter.

Dear KV,

Why do some modern network protocols not have sequence numbers? I would think that by now all protocol designers would have realized that having a simple sequence number in each packet helps people in debugging their network setups.

Out of Sequence

Dear OoS,

You might as well ask why people insist on not wearing seatbelts after all of the years that particular technology has been proven to save lives.

People will, it seems, persist in the optimistic belief that everything will be OK so long as they are otherwise careful. They think that bad things happen only to other people's protocols, or packets, but not to theirs. Hope springs eternal and dies in the cold, cold winter of experience.

I want to make two points in response to your plea for sanity in network protocol design. The first is that it's not just having a sequence number that is important, but how the sequence number is used is important as well. Consider the sequence number in TCP, which counts the bytes that have been communicated between two endpoints. When TCP was designed, the fastest network in common use was a 10-Mbps Ethernet LAN. Pay attention, that's an M, not a G—10 megabits per second. At 10 Mbps, transmitting 2^{32} bytes of data takes approximately 3,400 seconds, or just less than an hour, which is an eternity to a computer. On commodity 10-Gbps hardware available today, it takes 3.4 seconds to transmit the same data, meaning that the sequence space rolls over about every four seconds. If a packet is lost for more than four seconds, there is a nonzero probability that data on the connection will get munged. With hardware that will be available quite soon, the time will drop to 0.3 seconds for the sequence space to roll over.

None of this is to say that TCP was poorly designed (heck, at least it had a sequence number), but it is important for designers of modern protocols to understand the future proofing-vs.-space trade-offs when selecting a sequence number. If at some point TCP is extended, then the sequence number could be increased to 64 bits, which even at 100 Gbps would require 46 years to roll the number over. Any packet lost in the network that long will be quite lost indeed. When you choose a sequence number, consider what you're protecting. With TCP it is protecting all the bytes transmitted so that none is lost or reordered on delivery. With other protocols it might be necessary to count whole messages only so that the receiver can say that packet A arrived before packet B rather than worrying about every byte in the message.

The second point I would like to make is that timestamps are not good sequence numbers. While it is common to believe that time always moves forward, this is often not the case in computing. Many bugs crop up in dealing with time on computers, not the least of which is that different clocks on different computers often proceed at different paces.

This is why we have protocols such as NTP (Network Time Protocol) and PTP (Precision Time Protocol) to discipline our computer clocks. Alas, computers don't like to be disciplined, and even when running a time protocol, the clocks on two computers are always somewhat offset from each other, so running a time protocol does not solve this problem. Leaving aside the mind-bending relativity problems of computer timekeeping—and trust me, you really want to leave those aside—the fact remains that using the time on a computer as a packet sequence number is problematic. Incrementing a counter is easier, faster, and less error-prone than making sure that the timestamp you received is monotonically increasing. For the case of packet sequencing, simpler is better—and simpler is a counter.

To those who design or hope to design network protocols, please, I beg of you, do not skimp on the sequencing numbers. The bytes you save today will bite you on the ass tomorrow.

KV

4.5 Debugging the Network

> *Have you tried turning it off and back on again?*
>
> Too many to name

For the most part, most people who are taught to code are only taught how to debug their code, locally, on the screen in front of them, in a single instance, even though most systems now use some form of network either explicitly, because they knowingly make connections to remote services to do their work, or implicitly, because they run *in the cloud*, wherever that is.

Programmers who are used to integrated development environments, or even a moderately functional debugger, are shocked at the tools they have to use to debug networked code. The best known of these tools, `wireshark`, is a marvel with its single-minded determination to be able to display information on any known network protocol, even some that are barely used, and I reference that tool and its command-line equivalent, `tcpdump`, in my response to the next letter. For all that `wireshark` is a marvel, the idea that this is our best tool for debugging distributed systems is, frankly, terrifying. Koders are used to being able to set a break point in a program, evaluate the state of the program's variables, and even take actions under varying conditions as the program executes. For network protocols there are few, if any, commonly used tools that would measure up to a debugger from 20 to 30 years ago. The typical way that someone debugs a network problem is to collect as much network data as possible and then attempt to sift through it with `wireshark` or similar tools until they have an "aha!" moment while staring at a packet trace. Furthermore, there are few if any references that explain how you might debug a network problem, in part because a networking problem can happen in so many places. The problem can occur at any layer, from the hardware to the network interface, to the driver that runs the network interface, to the network protocol on top of the driver, to the transport protocol on top of the network protocol, to the application on top of the transport protocol. Most people tend to look first at the area in which they have the most understanding, so if you're an application developer, you look at your application, and if you can't find the problem, you blame one of the many layers beneath you.

Another complication in debugging distributed systems is that networks are frequently prone to partial failure, which often appears as a lowering of throughput or an unexplained increase in latency. In most of the rest of software, systems can and will fail stop. If your program dereferences a NULL pointer, on most systems the operating system will terminate your program with a nasty message about segmentation fault. A network protocol might drop every other packet on the floor and expect that *some other layer* will fix the problem for us. The host or router that dropped the packet might be thousands of miles away, and not under our control, and so we'll never know which device it is that's causing the poor performance; we will only see that we're missing a lot of packets on our end of the connection.

At the very high end of distributed systems you are debugging a system as vast as a planet, over which you have very little or no control, and you only have a keyhole to look through to see what's actually going wrong. Some attempt to bring this problem under control is the subject of the next letter.

4.5 Debugging the Network

Dear KV,

I posted a question on a mailing list recently about a networking problem and was asked if I had a `tcpdump`. The person who responded to my question—and to the whole list as well—seemed to think my lack of networking knowledge was some kind of affront to him. His response was pretty much a personal attack: If I couldn't be bothered to do the most basic types of debugging on my own, then I shouldn't expect much help from the list. Aside from the personal attack, what did he mean by this?

Dumped

Dear Dumped,

It is always interesting to me that when people study computer programming or software engineering they are taught to use the creative tools (editors to create code, compilers to take that code and turn it into an executable) but are rarely, if ever, taught how to debug a program. Debuggers are powerful tools, and once you learn to use one, you become a far more productive programmer because, face it, putting `printf()` or its immoral equivalent throughout your code is a really annoying way to find bugs. In many cases, especially those related to timing issues, adding print statements just leads to erroneous results. If the number of people who actually learn how to debug a program during their studies is small, the number who learn how to debug a networking problem is minuscule. I actually don't know anyone who was ever directly taught how to debug a networking problem.

Some people (the lucky ones) are eventually led to the program you mention, `tcpdump`, or its graphical equivalent, `wireshark`, but I've never seen anyone try to teach people to use these tools. One of the nice things about `tcpdump` and `wireshark` is that they're multiplatform, running on both Unix-like operating systems and Windows. In fact, writing a packet-capture program is relatively easy, as long as the operating system you're working with gives you the ability to tap into the networking code or driver at a low enough level to sniff packets.

Those of us who spend our days banging our heads against networking problems eventually learn how to use these tools, sort of in the way that early humans learned to cook flesh. Let's just say that though the results may have been edible, they were not winning any Michelin stars.

Using a packet-capture tool is, to a networking person, somewhat like using a thermometer is to a parent. It is likely that if you ever felt sick when you were a child, at least one of your parents would take your temperature. If they took you to the doctor, the doctor would also take your temperature. I once had my temperature taken for a broken ankle (crazy, yes, but that doctor gave the best prescriptions, so I just smiled blithely and let him have his fun). That aside, taking a child's temperature is the first thing on a parent's checklist for the question "Is my child sick?" What on Earth does this have to do with capturing packets?

By far the best tool for determining what is wrong with programs that use a network, or even the network itself, is the `tcpdump` tool. Why is that? Surely, in the now 40-plus years since packets were first transmitted across the original ARPANET, we have developed some better tools. The fact is we have not. When something in the network breaks, you want to be able to see the messages at as many layers as possible.

The other key component in debugging network problems is understanding the timing of what happens, which a good packet-capture program also records. Networks are perhaps the most nondeterministic components of any complex computing system. Finding out who did what to whom and when (another question parents often ask, usually after a fight among siblings) is extremely important.

All network protocols, and the programs that use them, have some sort of ordering that is important to their functioning. Did a message go missing? Did two or more messages arrive out of order at the destination? All of these questions can potentially be answered by using a packet sniffer to record network traffic, but only if you use it!

It's also important to record the network traffic as soon as you see the problem. Because of their nondeterministic nature, networks give rise to the worst types of timing bugs. Perhaps the bug happens only every so many hours, because of a rollover in a large counter; you really want to start recording the network traffic before the bug occurs, not after, because it may be many hours until the condition comes up again.

So, here are some very basic recommendations on using a packet sniffer to debug a network problem. First, get permission (yes, it really is KV giving you this advice). People get cranky if you record their network traffic, such as instant messages, e-mail, and banking transactions, and then post it to a mailing list. Just because some person in IT was dumb enough to give you root or admin rights on your desktop does not mean you should just record everything and send it off.

Next, record only as much information as you need to debug the problem. If you're new at this, you'll probably have the program suck up every packet so you don't miss anything, but that's problematic for two reasons: the first is the previously mentioned privacy issue; and the second is that if you record too much data, finding the bug will be like finding a needle in a haystack—only you've never seen a haystack that big. Recording an hour of Ethernet traffic on your LAN can capture a few hundred million packets. No matter how good a tool you have, it's going to do a much better job at finding a bug if you narrow down the search.

If you do record a lot of data, don't try to share it all as one huge chunk. See how these points follow each other? Most packet-capture programs have options to say, "Once the capture file is full, close it and start a new one." Limiting files to one megabyte is a nice start.

Finally, do not record your data on a network file system. There is no better way to ruin a whole set of packet-capture files than by having them capture themselves.

So there you have it: a brief introduction to capturing data so you can debug a networking problem. Perhaps now you can get yelled at on a mailing list for something more egregious than not taking your network's temperature before calling the doctor.

KV

4.6 Latency

> *When it absolutely, positively, has to be there overnight.*
>
> FedEx ad

Perhaps one of the most misunderstood parts of networked systems is the effect of latency on performance and system stability. I'd like to blame this on the fact that all network connections are sold or promoted based on the available bandwidth, but for people in technical fields this should not be an excuse. How long a message takes to get from point A to point B is an intrinsic part of how well the connection can be utilized as well as whether or not a mis-chosen timeout will result in an error. One of the things I keep in mind at all times is the scale of latency between various points, because this makes debugging broken networked systems easier. Admiral Grace Murray Hopper famously carried around a wire that was as long as a nanosecond so that she could explain these problems to nontechnical people, but I'd argue that we need to keep more in mind. On a quiet, properly functioning network, latencies are under a millisecond. If you're on a very fancy 10 Gbps or higher switch, they are probably closing in on tens of microseconds. Once you start moving around, the planet latency becomes more interesting. Latency across North America is about 70 ms. From Silicon Valley to Europe it's about 150 ms, and over to Asia it's about 110 ms. Keeping these rough numbers in mind when debugging latency problems will save you a lot of headaches in the long run.

4.6 Latency

Dear KV,

My company has a very large database with all of our customer information. The database is replicated to several locations around the world to improve performance locally so that when customers in Asia want to look at their data, they don't have to wait for it to come from the United States, where my company is based.

A few months ago the company upgraded its software, which required all of the records in the customer database to be updated as well. When we tested the upgrade program, it took only a few minutes to update a large number of records, but when we had to update the Asian customers, a process that had to run in Asia and that touched data in the United States, the process began to take a lot longer. Since the company has a very fast network connecting the U.S. and the Asian offices, it's hard to understand how the distance could matter. There must be another reason for the time it is taking to run these programs.

Baffled with Bandwidth

Dear Baffled,

There is a big difference between having a big pipe and knowing how to use it. Two things matter in networking: bandwidth and latency. Unfortunately, most people think only of the former, not the latter. Latency is the time a message takes to get from point A to point B (e.g., a client and a server). If I were to guess, I would figure that you tested the conversion program on a local network, probably 100 Mbit Ethernet, where latencies are generally less than 1 ms. You then ran the program remotely across your very fast network and found that it ran much more slowly.

I bet it ran 100 times as slow. How did I pick 100 times?

Easy, the average round-trip time across the Pacific is about 100 ms. What you forgot was a very important constant, and then you forgot how networks work.

The very fast network you speak of was probably sold on a bit-per-second basis, so maybe you have a 1 Gbps link between Japan and the United States, but that's a measure of bandwidth, not latency. It's the latency that matters in your case. Why? Because your conversion process very likely takes one record at a time, packs it up, makes a request to the server, and then waits for a response.

When the underlying network has very low latency, like a local network, this packing up of a single request and waiting for a response is barely noticed; but if you move to a higher-latency network, it crushes your system's performance under its iron heel.

The thing you forgot is c, the speed of light. Let's say you were running the conversion process in Tokyo and it was talking to a database in California. It's about 5,000 miles from Tokyo to California, and the speed of light is 186,000 miles per second, so a beam of light should be able to make it from Tokyo to California in about 0.027 seconds. That's 27 ms for the absolute fastest time between those two points, a round-trip of 54 ms. That's already a factor of 50 slower than your LAN.

Of course, packets don't travel point to point. They are stored and forwarded at various points along their journey; that's how the Internet works. Each waypoint (the technical term is router) introduces its own bit of delay to the packet's journey, until what we have is an average of 50 ms each way between Japan and California. Now you have a difference of a factor of 100 between your LAN and the real network. Reality bites, and in this case it bit you hard. If converting some number of records took five minutes locally, it's going to take 500 minutes (8-1/3 hours) remotely. If I were you, I would bring a book to work, or download a movie like the rest of your co-workers, before you start any more database conversions.

There are ways to ameliorate these problems, though getting around the speed of light has eluded better minds for quite a while. The first is to run all the conversions locally; the second is to write the conversion program so that it batches requests. This makes more efficient use of the high-bandwidth network that your company has probably paid a small fortune to rent. If the database conversion on the server side can process a batch of records in less time than it takes to move all those records across the link, then you will gain some time; but if each record must be processed serially, then you're always going to suffer with slow performance over a long-distance link.

KV

4.7 Long Distance Runaround

> *Never underestimate the bandwidth of a station wagon full of tapes hurtling down the highway.*
>
> Andrew S. Tanenbaum

A related mental problem to not appreciating latency is not understanding the difference between a stop-and-go protocol and a streaming protocol. Most people assume that the whole world is the Transmission Control Protocol (TCP), the best known and most frequently taught network protocol, but many distributed filesystems, such as the Network File System (NFS) and others, are block-based, stop-and-go protocols. Even when NFS is layered over TCP, it is still a stop-and-go protocol, although the transport layer is streaming. The next letter and response look at what happens when you mix high latency with a stop-and-go protocol, and the results, of course, are not good.

Dear KV,

I've been asked to optimize our NFS (network file system) set up for a global network, but NFS doesn't work the same over a long link as it does over a LAN. Management keeps yelling that we have a multigigabit link between our remote sites, but what our users experience when they try to access their files over the WAN link is truly frustrating. Is this just an impossible task?

Feeling Stretched Across the Sea

Dear Stretched,

The number of people who continue to confuse bandwidth with latency, and who don't seem to understand the limitations of the speed of light, does not seem to be decreasing, even though I kindly pointed this out in another context some time ago ['Latency and Livelocks,' ACM Queue, March/April 2008]. I would have thought that by now word would have gotten out that it doesn't matter how fat your pipe is, over a long distance latency is your biggest challenge. I suspect that this kind of problem is going to continue to come up, because since the tech bubble collapse in 2001, the number of cheap, long-distance fibers has not decreased. The world is awash in a sea of cheap bandwidth. Latency, on the other hand, is another story.

To understand why latency is killing your performance, it pays to understand how the NFS protocol works, at least at a basic level. NFS is a client/server protocol where the user's machine, the client, is trying to get files from the server. When the user wants a file, the client makes several requests to the server to get the data. Very simply speaking, the client has to look up the file—that is, tell the server which file it wants to read—and then it has to ask for each block of the file. The NFS protocol tries to read blocks from the file in 32 KB chunks, and it has to ask for each block in succession.

What does this have to do with latency? Many of the operations in NFS require that a previous operation has completed. Obviously the client cannot issue a READ request before it has looked up the file, and just as obviously it cannot issue a READ request for the next block in a file until it has received the previous one. Reading a file across NFS then looks like the following list of operations:

- Look up file.
- Read block 1.
- Read block 2.
- …
- Read block N.
- Done.

Between each of these steps the client has to wait for an answer from the server. The farther away the server is, the longer that response is going to take. NFS was originally designed to work in a LAN setting: one where computers were connected by a fast, 10-megabit (yes, you read that right, 10-megabit) network. The Ethernet LANs on which NFS was first deployed had round-trip latencies in the 5- to 10-millisecond range. During this same period computers had CPUs that were measured in the tens of megahertz (yes, again, go back and read that, tens of megahertz). The best thing you could say about this arrangement was that the network was far faster than the CPU, so users didn't mind waiting on the file server because they were used to waiting. This was so long ago that people still smoked cigarettes, and processing a long file generally meant it was time for a smoke break.

In the local area, speeds have continued to improve, both in bandwidth and latency. Most users now have 1-gigabit links to their networks, and LAN latencies are in the submillisecond range. Unfortunately, the speed of light gets involved when you start creating networks over global distances. It's typical for a trans-Pacific network link to have a 120-ms round-trip time. Latencies across the Atlantic and North America are lower, but by no means are they fast, and they're not going to get much faster unless someone finds a way to violate some important parts of Einstein's theories. Every physicist wants to violate Einstein, but thus far the great man has remained pretty chaste.

Look at it this way: for every mile between the client and the server, a message cannot get to the server and back to the client in less than 10 microseconds, because light travels one mile in 5.4 microseconds in a vacuum. In a fiber-optic network, or in a copper cable, the signal travels considerably slower. If your server is 1,000 miles from your client, then the best round-trip time you could possibly achieve is 10 milliseconds.

Let's pretend for a moment that you happen to have an ultra-high-tech, light-in-a-vacuum network, and that your round-trips are always 10 ms. Let's also pretend that bandwidth is not a problem. How long will it take to read a 1 MB file over that perfect link? If each request is for 32 KB, 32 requests will be sent, which works out to 320 milliseconds. Not so bad, you think, but people notice computer lags of just 200 ms. Whenever your users open a file, they're going to experience this lag, and they're going to be standing in your doorway, if you have a doorway, whining about how your expensive network link is just too slow. They're not going to like the best answer, which is, "Do not use NFS over long distances," but that truly is the best answer.

There is one protocol that has been endlessly optimized over the past 30 years to deal with remote files over distances of more than a mile, and that's TCP. "Wait! I use NFS over TCP!" I hear you cry. That may be, but once you layer NFS on top of TCP, you have already lost; because of the block nature of NFS just described, you will never be able to use the underlying TCP connection efficiently. Only if NFS were able to get whole files from the server in one request would it be able to start using the underlying protocol efficiently.

There are things that can be done to improve your situation. While it's unlikely you'll be able to do much to tune NFS to do the right things, you can tune your underlying TCP settings; this is normally done on a system-by-system basis, however, which means that you might sacrifice some local performance to improve the user's remote experience. Search the web for information on "tuning TCP for high bandwidth/delay product networks" and apply their suggestions.

Remember to test what you try, instead of just blindly applying the numbers you are given. By tuning TCP, it's quite easy to make things worse than they were in the default case. I suggest using a program such as `scp` to copy a file that you're also trying to copy across NFS and compare the times. I know that `scp` has some cryptography overhead, but suggesting that people use `rcp` is like suggesting that they learn to juggle by starting with scissors.

I've included a link to a decent bandwidth/delay calculator, just to get you started: http://www.speedguide.net/bdp.php.

KV

4.8 The Network Is the Computer

> ...and the computer is down.
>
> Anon

Failing at scale is a problem that has come to us through the advent of inexpensive hardware and the proliferation of networked systems. The challenge is always that it's hard to know when a scale failure will crop up, and simulations have a very poor track record of helping us to avoid these problems in the real world. Many implementors have therefore chosen a *push it until it breaks* mentality that works well, until the system breaks. One of the major challenges to this push-it mentality is that when a distributed system fails, it is usually a cascading failure, with many parts failing and causing other parts to fail in turn, and tracking down the root cause then becomes a reason for anything from heavy drinking to suicidal ideation.

One way in which developers of large networked systems try to maintain their limited sanity is to deploy two networks, one for internal control traffic and the other for customer-facing traffic. Having a split architecture means that it might be possible to use the control network to see what's wrong when the customer network goes nuts or is subjected to a denial-of-service attack. A split architecture is also necessary in a test network to keep control traffic from interfering with the tests and vice versa.

How one might go about debugging a system that fails at scale should be the topic of at least one good book, but for now we'll have to be content with the next letter and its answer.

Dear KV,

Our project has been rolling out a well-known, distributed key/value store onto our infrastructure, and we've been surprised more than once when a simple increase in the number of clients has not only slowed things, but brought them to a complete halt. This then results in rollback while several of us scour the online forums to figure out if anyone else has seen the same problem. The entire reason for using this project's software is to increase the scale of a large system, so I have been surprised at how many times a small increase in load has led to a complete failure. Is there something about scaling systems that's so difficult that these systems become fragile, even at a modest scale?

Scaled Back

Dear Scaled,

If someone tells you that scaling out a distributed system is easy, they are either lying or drunk, and possibly both. Anyone who has worked with distributed systems for more than a week should have this knowledge integrated into how they think, and if not, they really should start digging ditches. Not to say that ditch digging is easier, but it does give you a nice, focused task that's achievable in a linear way, based on the amount of work that you put into it. Distributed systems, on the other hand, react to increases in offered load in what can only politely be referred to as nondeterministic ways. If you think programming a single system is hard, programming a distributed system is a nightmare of Orwellian proportions where you almost are forced to eat rats if you want to join the party.

Nondistributed systems fail in much more predictable ways. Tax a single system and you run out of memory or CPU or disk space or some other resource, and the system has little more than a snowball's chance of surviving a Hawaiian holiday. The parts of the problem are so much closer together and the communication between those components is so much more reliable that figuring out "who did what to whom" is tractable. Unpredictable things can happen when you overload a single computer, but you generally have complete control over all of the resources involved. Run out of RAM? Buy more. Run out of CPU, profile and fix your code. Too much data on disk? Buy a bigger one. Moore's law is still on your side in many cases, giving you double the resources every 18 months.

The problem is that eventually you will probably want a set of computers to implement your target system. Once you go from one computer to two, it's like going from a single child to two children. To paraphrase an old comedy sketch, if you only have one child, it's not the same has having two or more children. Why? Because when you have one child and all the cookies are gone from the cookie jar, *you know who did it!* Once you have two or more children, each has some level of plausible deniability. They can, and will, lie to get away with having eaten the cookies. Short of slipping your kids a truth serum at breakfast every morning, you have no idea who is telling the truth and who is lying. The problem of truthfulness in communication has been heavily studied in computer science, and yet we still do not have completely reliable ways to build large distributed systems.

One way that builders of distributed systems have tried to address this problem is to put in somewhat arbitrary limits to prevent the system from ever getting too large and unwieldy. The distributed key store Redis had a limit of 10,000 clients that could connect to the system. Why 10,000? No clue, it's not even a typical power of 2. One might have expected 8,192 or 16,384, but that's probably another article. Perhaps the authors had been reading the Tao Te Ching and felt that their universe only needed to contain 10,000 things. Whatever the reason, this seemed like a good idea at the time.

Of course, limiting the number of clients is only one way of protecting a distributed system against overload. What happens when a distributed system moves from running on 1 Gbps network hardware to 10 Gbps NICs? Moving from 1 Gbps to 10 Gbps doesn't "just" increase the bandwidth by an order of magnitude, it also reduces the request latency. Can a system with 10,000 nodes move smoothly from 1 G to 10 G? Good question; you'd need to test or model that, but it's pretty likely that a single limitation such as number of clients is going to be insufficient to prevent the system from getting into some very odd situations. Depending on how the overall system decides to parcel out work, you might wind up with hot spots, places where a bunch of requests all get directed to a single resource, effectively creating what looks like a denial-of-service attack and destroying a node's effective throughput. The system will then fail out that node and redistribute the work again, perhaps picking another target, and taking it out of the system because it looks like it, too, has failed. In the worst case, this continues until the entire system is brought to its knees and fails to make any progress on solving the original problem that was set for it.

Distributed systems that use a hash function to parcel out work are often dogged by this problem. One way to judge a hash function is by how well-distributed the results of the hashing function are, based on the input. A good hash function for distributing work would parcel out work completely evenly to all nodes based on the input, but having a good hash function isn't always good enough. You might have a great hash function, but feed it poor data. If the source data fed into the hash function doesn't have sufficient diversity (that is, it is relatively static over some measure, such as requests), then it doesn't matter how good the function is, as it still won't distribute work evenly over the nodes.

Take, for example, the traditional networking 4-tuple, source and destination IP address, and source and destination port. Together this is 96 bits of data, which seems like a reasonable amount of data to feed the hashing function. In a typical networking cluster, the network will be one of the three well-known RFC 1918 addresses (192.168.0.0/16, 172.16.0.0/12, or 10.0.0.0/8). Let's imagine a network of 8,192 hosts, because I happen to like powers of 2. Ignoring subnetting completely, we assign all 8,192 hosts addresses from the 192.168.0.0 space, numbering them consecutively 192.168.0.1–192.168.32.1. The service being requested has a constant destination port number (e.g., 6379), and the source port is ephemeral. The data we now put into our hash function are the two IPs and the ports. The source port is pseudo-randomly chosen by the system at connection time from a range of nearly 16 bits. It's nearly 16 bits because some parts of the port range are reserved for privileged programs, and we're building an underprivileged system. The destination port is constant, so we remove 16 bits of change from the input to the function. Those nice fat IPv4 addresses that should be giving us 64 bits of data to hash on actually only give us 13 bits, because that's all we need to encode 8,192 hosts. The input to our hashing function isn't 96 bits but is actually fewer than 42. Knowing that, you might pick a different hash function or change the inputs, inputs that really do lead to the output being spaced evenly over our hosts. How work is spread over the set of hosts in a distributed system is one of the main keys to whether that system can scale predictably, or at all.

An exhaustive discussion of how to scale distributed systems is a topic for a book far longer than this piece, but we can't leave the topic until we talk about what debugging features exist in the distributed system. "The system is slow" is a poor bug report; in fact, it is useless. However, it is the one most often uttered in relation to distributed systems. Typically the first thing that users of the system notice is that the response time has increased and that the results they get from the system take far longer than normal. A distributed system needs to express, in some way, its local and remote service times so that the systems operators, such as the DevOps or systems administration teams, can track down the problem. Hot spots can be found through the periodic logging of the service request arrival and completion on each host. Such logging needs to be lightweight and not directed to a single host, which is a common mistake. When your system gets busy and the logging output starts taking out the servers, that's bad. Recording system-level metrics, including CPU, memory, and network utilization will also help in tracking down problems, as will the recording of network errors. If the underlying communications medium becomes overloaded, this may not show up on a single host, but will result in a distributed set of errors, with a small number at each node, which will lead to chaotic effects over the whole system. Visibility leads to debuggability; you cannot have the latter without the former.

Coming back around to your original point, I am not surprised that small increases in offered load are causing your distributed system to fail, and, in fact, I am most surprised that some distributed systems work at all. Making the load, hot spots, and errors visible over the system may help you track down the problem and continue to scale it out even further. Or, you may find that there are limits to the design of the system you are using, and you'll have to either choose another or write your own. I think you can see now why you might want to avoid the latter at all costs.

KV

4.9 Failure to Scale

Communication is key.

— Anon

One of the most common reasons that distributed systems fail to scale isn't stupidity, although that's KV's usual go-to for an explanation, but sometimes it's simple naivety. A naive first attempt at a job control system, for want of a better term, is given in the next call and response. Like multithreaded programming, programming with nonblocking IO is something that every koder who works on network kode needs to learn, the sooner the better.

Dear KV,

I have been digging into a network-based logging system at work because, from time to time, the system jams up, even when there seems to be no good reason for it to do so. What I found would be funny, if only it weren't my job to fix it: the central dispatcher for the entire logging system is a simple `for` loop around a pair of read and write calls; the `for` loop takes input from one of a set of file descriptors and sends output to one of another set of file descriptors. The system works fine so long as none of the remote readers or writers ever blocks, and normally that's not a problem. The problem has come about because what was once handling fewer than 10 machines is now handling 40, some of which are remote across a wide area network. The obvious fix is to make the code nonblocking, but what I'm surprised about is that anyone would write code this way. It's obvious from the first time you look at the code that it cannot scale.

Blocked and Loopy

Dear Loopy,

I would like to say that I'm sure the original author of the code you're looking at wasn't trying to torture you; but after seeing many similar pieces of code, it's hard for me to continue to accept this particular bit of make-believe. What you're probably looking at is "throwaway" or "prototype" code that got away. The schlimazel who wrote the code probably had a boss pop into his cube one day with a "great idea" to improve the logging system by using the network and a central dispatcher, and then asked the programmer to code up something simple to toss around. That something simple is what you now see. In my mind, I see the programmer getting the code running and, since programmers are optimists, being excited when it ran and considering it done.

The next thing I see is that once the code was deployed, people found a use for it. Code that people don't find a use for rarely causes problems, because it rarely gets executed. From 10 clients, it went to 20, and then on to the point where it broke and someone asked you to look at it.

If I were you, I'd count your blessings. Taking a single, simple read/write loop and converting it into a reasonably robust, nonblocking piece of code, while not trivial, isn't a massive undertaking. Of course, while you're at it, you're going to add code to report when your clients are slow or disconnect or cause problems, right? Right! You could easily spend days hacking around and polishing a system like this, but I would suggest that you just add enough code and hooks so that when the system goes to 100 nodes, you can split your dispatcher and run more than one of them simultaneously on separate nodes, because that's the next thing you'll have to do for scalability. If you don't do this correctly, then your successor will be writing me a letter exactly like this one.

KV

4.10 Port Squatting

> *Bureaucracies are designed to perform public business. But as soon as a bureaucracy is established, it develops an autonomous spiritual life and comes to regard the public as its enemy.*
>
> Brooks Atkinson

It may come as a surprise to my more regular readers, but when it comes to the use and abuse of network technology, KV tends to, at least at first, side with the powers that be, in part because for a good deal of time those powers were not institutional but were constructed along what someone with a more political bent would call anarchist lines. The IETF, the most respected of the networking standards bodies and the creators of the Internet protocols, has always emphasized, "rough consensus and running code," which is the exact right tack to take with most technologists. As the Internet protocols took off and the bodies started by the IETF grew into institutions, though, many of them have become ossified and bureaucratic, so this letter and response cover both the respect KV has for these institutions and a call for their reform. "The tree of liberty must occasionally be watered with the blood of patriots," according to Jefferson, and though that may seem a bit bloody in the literal sense, as a metaphor for change in large organizations it remains apt.

Dear KV,

A few years ago you upbraided some developers for not following the correct process when requesting a reserved network port from IETF (Internet Engineering Task Force). While I get that squatting a used port is poor practice, I wonder if you, yourself, have ever tried to get IETF to allocate a port. We recently went through this with a new protocol on an open source project, and it was a nontrivial and frustrating exercise. While I wouldn't encourage your readers to squat ports, I can see why they might just look for unallocated ports on their own and simply start using those, with the expectation that if their protocols proved popular, they would be granted the allocations later.

Frankly Frustrated

Dear Frankly,

Funny you should ask this question at this point. This summer I, too, requested not one but two ports for a service I'd been working on (Conductor: `https://github.com/gvnn3/conductor`). I've always been annoyed that there isn't a simple, distributed, automation system for orchestrating network tests, so I sat down and wrote one. The easiest way to implement the system was to have two reserved ports, one for the conductor and one for the players so that each could contact the others independently without have to pass ephemeral ports around after they were allocated by the operating system during process startup.

Simple enough, you might think. It's not actually IETF to which one applies; it's IANA (Internet Assigned Numbers Authority). It has a form you fill out on its web site, detailing your request (`https://www.iana.org/form/ports-services`), which asks fairly reasonable questions about who you are, which transport protocol your protocol uses (UDP, TCP, SCTP, etc.), and how the protocol is used. Because there are only 16 bits in the port field for UDP, TCP, and SCTP, space is limited, so you can see why IANA would want to be careful in its port allocations. Looking over the current assignments, we can see that nearly 10 percent of the space has already been allocated for TCP, with more than 6,100 assigned ports for TCP.

I submitted my request for a pair of ports over TCP and SCTP on July 7. I applied for both because it made sense to address both of the currently available, reliable, transport protocols. As I write this, it is September 6, and I'm assured that by September 8 I'll have a single port assigned. Let's look at the process.

Once you submit your port request, it goes into a ticketing system, RT (request tracker), which is looked after by someone whom I'll call a secretary. The secretary seems to do some form of triage on the ticket and then passes it along to someone else. For the past two months, the secretary asked clarifying questions about the use of the two port numbers. It was plain from the interaction that the secretary did not have any significant networking knowledge but acted as a pass-through for the experts reviewing the case.

As might be expected with any sort of overly bureaucratic process and with this form of telephone game, information was often lost or duplicated, requiring me to explain at length how I was going to use the ports. In the end, I was contacted by an expert (someone actually knowledgeable about networking technology), and we agreed that the service could be built with one port number. I say "agreed," but mostly I relented, because I was going to do this right, even if I put my fist through a whiteboard, and let me tell you, I came very close to doing just that.

This brings me to a few statistics about the assigned numbers. Many of the assignments for TCP aren't for a single port, but for multiple ports, meaning the number of services is fewer than the 6,100-plus assigned ports. Not only are there many services with more than one port, but it would also seem that dead assignments are not garbage collected, which means that although only 10 percent of the space is used, there is no way to reclaim ports when protocols or the companies that created them die. Looking through the list of assigned ports is a walk down the memory lane of failed companies.

All of which is not to encourage people to squat numbers, but it is pretty clear that IANA could do some work to streamline the process, as well as reclaim some of the used space. The biggest problem actually exists in the first 1,024 ports, which most operating systems consider to be "system" ports. A system port is usable only by a service running as root, and this is considered privileged. The domain name system, for example, runs on port 53. It's in this low space that IANA needs to get its collective act together and kill off a few services. Although I'm sure all of you are using port 222 Berkeley rshd with SPX auth each and every day.

KV

4.11 Networking in the Raw

> *Everything is derivative.*
> *Everything is a remix, and we all*
> *stand on the shoulders of giants.*
>
> Alexis Ohanian

Everyone wants a clean slate on which to build their system, in part because it feels more satisfying and in part because most technologists are control freaks who are hyped up on hubris. There is nothing inherently wrong with having such yearnings, but they must be tempered by good judgment.

Good judgment in this case might include looking to protocols other than the usual suspects, IP, UDP, and TCP. There are many standard and near standard protocols that attempt to work around the known deficiencies in the big three for various cases, and that's where one should start, before breaking out a nice sheet of paper. It comes down to doing your homework, which to some is boring, but the truly gifted practitioner of the art knows that doing so will shorten their development time and produce a better result.

The next letter and response cover this situation, the one in which the uninitiated, or insufficiently initiated, wish to throw the baby out with the bathwater because they want a clean tub to bathe in.

Dear KV,

The company I work for has decided to use a wireless network link to reduce latency, at least when the weather between the stations is good. It seems to me that for transmission over lossy wireless links we'll want our own transport protocol that sits directly on top of whatever the radio provides, instead of wasting bits on IP and TCP or UDP headers, which, for a point-to-point network, aren't really useful.

Raw Networking

Dear Raw,

I completely agree that the best way to roll out a new networking service is to ignore 30 years of research in the area. Good luck.

Second only to operating system developers (all of whom want to rewrite the scheduler[2]) are the networking engineers and developers who want to write their own protocol. "If only we could go at it with a clean sheet of paper, we could do so much better than the ARPANET, since that was designed for old, crappy hardware, and ours is shiny and new." That statement is both true and false, and you had better be damned sure about which side of the Boolean logic your idea lies on before you write a single line of new code.

The Internet protocols are not the be-all and end-all of networking, but they have had more research and testing time applied to them than any other network protocols currently in existence. You say you're building a wireless network with I'm sure the highest quality gear you can buy. Wireless networks are notoriously lossy, at least in comparison to wired networks. And it turns out that there has been a lot of research done on TCP in lossy environments. So although you will pay an extra 40 bytes per packet to transport data over TCP, you might get some benefit from the work done to tune the bandwidth and round-trip-time estimators that will exist in the nodes sending and receiving the data.

Your network is point-to-point, which means you don't think you care about routing. But unless all the work is always going to be carried out at one or the other end of this link, you're eventually going to have to worry about addressing and routing. It turns out that someone thought about those problems, and they implemented their ideas in, yes, the Internet protocols.

The TCP/IP protocols aren't just a set of standard headers; they are an entire system of addressing, routing, congestion control, and error detection that has been built upon for 30 years and improved so that users can access the network from the poorest and most remote corners of the network, where bandwidth is still measured in kilobits and latencies exceed half a second. Unless you're building a system that will never grow and never

---

2. George Neville-Neil: Bugs and Bragging Rights, in: Queue 11.10 (Oct. 2013), 10–12, URL: https://doi.org/10.1145/2542661.2542663.

be connected to anything else, you had better consider whether or not you need the features of TCP/IP.

I am all for clean-sheet research into networking protocols. There are many things that have not been tried and some that have been but didn't work at the time. Your letter implied not so much research, but rollout, and unless you've done your homework, this type of rollout will flatten you and your project.

KV

4.12 Pointless PKI

> *Relying on the government to protect your privacy is like asking a peeping tom to install your window blinds.*
>
> John Perry Barlow

There are many places where networking and security intersect, but the place where they collide is when building Public Key Infrastructure. An easy to build, maintain, and understand PKI system remains a dream in the networking and security communities. Many papers, protocols, and pieces of code have been built and pushed into the public sphere to try to solve this problem, and yet the problem persists. The reason this problem persists is that the definition of a secure infrastructure is slippery and not agreed upon by all the possible participants. Most users barely notice security issues until they're violated, which we turn to in Section 5.11, while the owners of most infrastructure want the system to only be secure for themselves, and not for their users, because a system with good security would not allow them to exploit, I mean monetize, their users. Governments, even more than private companies, do not want to see truly secure systems, for then how could they effectively spy on, I mean protect and serve, their citizens? With this many powerful interests arrayed against the idea of security in distributed systems, it's amazing that any good work ever gets done in this area.

Dear KV,

I work at a large web-based company, and we're looking at a way to secure our traffic. Unlike most companies, this is not to secure the traffic between remote offices but actually to secure all the traffic inside the company, between the front-end web servers and our back-end databases. We've had problems in the past with internal compromises, and management has decided that the only way to protect the information is to encrypt it during transmission. We won't be storing the data encrypted because it's too hard to get everyone to rewrite their applications. Building a system like this is no easy feat because we have thousands of servers involved in making our systems work. I'm building a PKI system to handle all the keys necessary, one for each service we provide, and am wondering if you have any advice on how to secure data inside a company as opposed to making the service itself secure.

Keyed Up over Security

Dear Keyed,

Yes, you're right, building such a system is no easy feat, and the worst part is, even if you succeed, it will be completely pointless. There are several things that confuse me in your letter, so I'll try to address them logically, which is more than I can say for your management, who seem to be in what is kindly referred to as "reactive mode" and what I would term "putting their heads in your sand" or somewhere else.

By internal compromises I suspect you mean that some employees have been making off with your data. Internal compromises and leaks are, of course, always a risk at any company, and the larger the company the bigger the risk. The more people you have involved in a group effort, the more likely you're going to wind up with a few people who you should not have hired. How do you keep these infernal internal people from doing things with the data that they should not? I can tell you that encrypting all traffic is not going to help very much. It is very unlikely that an inside attacker would put a packet sniffer on the network, collect a day or week's worth of data, and then walk out with it. Trying to sift through that much information is far too much work, and besides, you've provided them with a much easier target.

If your company is storing the data in a back-end database, then that is the place where the internal attacker would go to get their data. Why sift through packet traces when a few SQL statements and a DVD burner, or fast net connection, would provide you with much better data? If you're storing sensitive data, then it is the data that needs to be secured, not the network! What if the attacker walks off with the backups, as has happened in several cases recently? If the data in the database is not secured, that is, hashed or encrypted, then he who has the backups has the data.

Another important concern that most people don't understand in this sort of system is the concept of "need to know." Governments and the military, which are just two sides of the same coin, attempt to set up systems such that sensitive data is only seen or modified by people who actually need to work with that data, hence, "need to know." Databases and other computer systems can also be set up in similar ways, such that only the small number of people who need to work with any particular bits of data actually work on them. Honest people won't care that they don't have access to all the data because they have what they need to do their jobs, and the dishonest ones will have access to less data, thereby reducing the chance of a compromise.

A frequent mistake in setting up secure systems is to encrypt everything. If you encrypt all data, then everyone has to have keys, and the keys can be lost or stolen thereby leading to a compromise. Encrypt only what you must encrypt to secure your business; then only a small number of people will need keys or access to the sensitive data.

Finally, you don't say anything about auditing in your mail. The best way to find the dishonest people in your system is not by encrypting all the communication and hoping that that keeps them out but by auditing when sensitive data is read, modified, or deleted. Keeping a log of "who did what to whom" and reviewing that log on a regular basis is the best way to find the abusers of your system.

I'm sorry I didn't tell you how to implement a good PKI system; it's a fascinating topic, but it's not one that is going to help you at all, except in making your corporate masters feel more secure, when in reality they won't be.

KV

4.13 Standard on Standards

> *The nice thing about standards is that there are so many to choose from; furthermore if you do not like any of them, you can just wait for next year's model.*
>
> Rear Adm. Grace Murray Hopper

No other area of computing is more defined by standards than that of networking, which makes a great deal of sense for many reasons. The only way to convince two or more heterogenous computing systems to correctly exchange information is by defining some common form of exchange between them. Human beings do this all the time with language, but human languages are organic, illogical, and error prone, all of which make them a very poor model for computer-to-computer communication. Like programming languages, which, in the best case, are built up out of logical pieces that represent the common underlying features of computing systems, such as boolean logic, stored programs, and addressable storage (aka memory), network protocols should be defined in terms that can be easily understood and implemented using the available metaphors in networking.

The goal of any form of communication is to take a concept from point A and represent it identically at point B. In human language we wish to take the concept that we have in our head and, by speech or writing, cause the receiver of that message to have that same concept. Network protocols have a much more narrow definition of *concept*, as the goal of all network communication is to take some piece of data, from a single bit to a group of many petabytes, and to transfer the data, in order, without duplications, and without losing it, from point A to point B. While this definition sounds simple when compared to human communication, there are many design constraints that have to be taken into account when designing a network protocol for the real world. KV has laid out many of these challenges in the sections leading up to this ultimate discussion of the chapter.

How you take an existing standard and turn it into code is the subject of this next letter and response, which elucidates a different set of concerns. The protocol designers can be thought of as architects, an unfortunately abused term in the computing world today, and the implementor is the construction team that takes the plans and builds the building. There are things that the architects don't know, or perhaps appreciate, in the building of a building, but which the construction team must know, or else the entire enterprise might fail. It is to those who construct these modern, invisible edifices that this letter is dedicated.

Dear KV,

I've been implementing a network protocol for my employer, and although I've heard people complain about technical specifications before, the group that designed this one must be particularly special. Not only is the text nearly impenetrable, but I also keep finding that they have left out important points, such as whether or not some fields are supposed to exist in some cases. I feel as if I have no choice but to find another implementation and test mine against it so that I know if it will actually work, a step I was going to take later in the development process. How can anyone be expected to implement software based on such a document?

Duped by Documentation

Dear Duped,

You have a document?! Consider yourself lucky! Actually, perhaps you're not really lucky. It turns out that the quality of standards varies just as widely as the quality of code, and a good way to find this out is to spend a couple of decades reading them, as KV has. If this is your first time implementing something from a "standard," you probably expected it to be written by intelligent professionals who were focused solely on making sure that the people implementing their ideas were able to do so quickly and efficiently with the least amount of ambiguity. What you probably didn't know was that such philosopher kings are the things of myth. Standards are written for many different reasons, some of them having nothing to do with the quality of the eventual product. There are even cases— and I know you will be shocked to hear this—where companies specifically send people to work on standards so that they will either never see the light of day or, when they do, will be unimplementable, therefore giving that company a commercial edge. Of course, my telling you that people are stupid, vain, and vile and that companies are just as likely to destroy innovation as to promote it doesn't really help you, but it does feel good.

On a slightly more practical level, there are several things to note when implementing a standard in code. The first is that rather than go directly to interoperability testing, which you allude to, you should start by marking up the standard. Now that specs are usually issued in PDF, there are several good programs that allow you to keep arbitrary notes with the document. KV actually prefers the pen and paper method, although for some standards, carrying around a printed copy can be cumbersome. Whatever your markup tool of choice, go somewhere quiet, sit down, and read the entire spec and make notes. Call out every ambiguity you find. If your notes runneth over, keep a separate file of them somewhere. Think of this as being similar to writing comments in your code. Some standards and specs have commentary, and some of this commentary is even useful, but often these documents arrive more as pronouncements from on high, though perhaps with fewer thous and thees in them.

Once you have marked up the document, the next thing to do is write tests for all of the cases that you can tease out of the document. I know, I often go on about how important testing is, but in this case it's truer than any other I can think of. I personally had the displeasure of working with a networking standard in which the authors had not consistently declared their padding bytes. In some places they very dutifully said, "These bytes are always 0," but in others they said nothing. It was only after I had written several tests for this protocol that I determined that they had meant every one of their declared fields to start on a 32-bit boundary. Once I saw how the bytes would look on the wire, something else made nearly impossible by the horrible notation used in the standard, their original intent became more obvious. Most people call this an "aha" moment, or, if they're in the bath, they yell, "Eureka!" I, and the people who sit near me at work, can tell you that "Eureka!" was not what I yelled. Let's just say that my dad was a sailor, and I curse like one. There is no way I would have been able to figure out this problem without my own test code.

We now come to one of the points you made in your letter: testing against other known implementations. If you're lucky enough to not be the first poor sap implementing the spec, then yes, interoperability testing might help you. I say might help because the person or group that implemented the code you're testing against may have been more confused than you are, so making your code work against theirs just means there are two interoperable, but also flawed, implementations of the same standard. Hurray for that. Do not assume that just because a version of the standard you're implementing exists that it is any good. The world is littered with systems that are interoperable but that are also wrong. I am thinking here of the many cases of network clients that have to work with code from a large company in the Northwest. I don't normally go after a particular vendor in this column, but it has been my experience that one particular vendor has been the source of more crap networking code than any other I have come across, so enough said.

One last recommendation is that you specifically call out which section of a standard or spec is being implemented in the code. For example:

```
/*
 * Update the Older Version Querier Present timers for a link.
 * See Section 7.2.1 of RFC 3376.
 */
```

and

```
/*
 * RFC 1122, Sections 3.2.2.1 and 4.2.3.9.
 * Treat subcodes 2,3 as immediate RST
 */
```

Both of these examples come from the TCP/IP stack implemented in FreeBSD, but this is a common practice when implementing code directly from a standard or specification, and it helps to improve your life in a few ways. First, writing things down is one way in which people reason about problems. So long as things are only voices in your head, they don't have the same concreteness as they do on paper or in a file. Once something is out of your head and you can examine it more objectively, you can do a better job of reasoning about whether or not what you thought was actually the case.

Second, these act, as do all good comments, as signposts to people who will maintain the code. There is nothing more frustrating than looking at some obscure piece of a function and wondering, "Now why did they do that?"—particularly if what was done doesn't make immediate sense. The code probably exists for a good reason, but it's important to separate the original programmer's capriciousness from the capriciousness of the standard. If it's a part of the standard, then it might look irrational, but for the sake of interoperability you'll have to leave it alone.

Of course, as you can imagine, I have some advice for standards writers as well. Perhaps the most important thing anyone working on a standard or spec can do is to be consistent in language and representation. At this point most of the constructs that would go into a new standard already exist, so please, stop inventing new ways to represent data structures. I'm sorry if you find it exciting to come up with new ways to represent bytes and bits on paper, but standards are not works of visual art; they need to be works of clarity. I happen to prefer the textual representation found in most RFCs, where you get labeled boxes, at most 32 bits wide. These are not the be-all and end-all of visual representation, but they're a damned good start, so please, start there.

Now, once you think you have a clearly written document, hand it to someone who is not working on your project and see if it really is clear. The idea that a group of people working closely on a standard are the right people to check the standard for clarity is ludicrous. After only cursory exposure to the set of ideas in the standard, an internal reviewer's brain will begin to fill in the blanks, and that's absolutely not what you want. You want someone who will call out the blanks and tell you where they are. Finally, have someone try to implement the spec—again, someone external to the group writing it—and then *listen to what they say!* Far too many times I've asked someone, "Did anyone review this?" and they said, "Yes, of course!" in a shocked tone as if I'd asked if they had showered that day and was impugning their sense of hygiene. And then when I asked, "Did you integrate their feedback?" they began to look quite sheepish.

Implementing a standard isn't too different from any other sort of implementation. The authors of something marked as a standard are not gods, and their utterances should not be taken as commandments. The short answer to your question is take notes, write tests, and keep a bottle of your favorite sedative nearby, because if you don't need it this time, well, you will eventually.

KV

5

No matter what the problem is, it's always a people problem.

Gerald M. Weinberg

Human to Human

Many people who enter technological fields, and this is perhaps most prevalent in the computing field, feel far more comfortable "talking" to machines than to other people. Now this isn't talking in the way we now talk to our ridiculously limited digital assistants, "Hey, Siri, buy me a beer." but talking in the way that programmers like to program, which is a form of communication, just not one that is generally recognized outside of the computing profession. Unfortunately, life is not always so simple and, more often than not, we find ourselves having to communicate with actual people. Messy, inexact, demanding, annoying, aggravating people. The success of a project often depends on our more human abilities, limited though they may be, compared to those who went "the other way" at university and studied the sciences or arts that related more to humans than to machines. This chapter presents the letters that came to KV but did not relate to code at all; they related to that problem that exists between the keyboard and the chair.

5.1 Of Pride and...

Pride goeth before destruction,
and a haughty spirit before a fall.

King James Bible, Book of Proverbs, 16:18

It is often said that "those who forget history are doomed to repeat it" and so whenever there is a chance to learn from history, even if that history has nothing, ostensibly, to do with software, KV tries to pay attention to what history has to say.

The story of the *Vasa* holds a special place in my heart because it has so many lessons packed into a very small, neat package, and history is many things, but it is rarely neatly packaged. While those of us who work in technology think that the problems of dealing with management, and management's lack of understanding of technology, developed as part of industrialization and specialization starting in the 18th and accelerating through the 19th, 20th, and 21st centuries, the story told by the *Vasa* shows us that these issues definitely pre-date modern industrial society. This fact should come as no surprise, because the problem here isn't technological, it is very human, but let's let the letter tell us the story.

Dear KV,

I teach computer science to undergraduate students at a school in California, and one of my friends in the English department, of all places, made an interesting comment to me the other day. He wanted to know if my students had ever read *Frankenstein* and if I felt it would make them better engineers. I asked him why he thought I should assign this book, and he said he felt that a book could change the way in which people think about their relationship to the world, and in particular to technology. He wasn't being condescending; he was dead serious. Given the number of *Frankenstein*-like projects that seem to get built with information technology, perhaps it's not a bad idea to teach these lessons to computer science undergraduates to give them some notion that they have a social responsibility. Do you agree?

CS Prof

Dear CS,

While I have to agree in general with the idea that telling and retelling stories is a good way to teach people, I have to say that the idea of using Mary Shelley's novel for this is very much antiquated and unlikely to be effective in a computer science class. I, myself, was once forced through a "Computers and Society" course in college, and although we didn't read *Frankenstein*, we were beaten over the head with a litany of how bad computers and technology were for society from a professor who was trivial to manipulate. All I had to do was agree with her every utterance and write technology-bashing essays for her class to get an A. Was this an effective use of time?

Of course not, it was a show. If you really want to reach an audience, you have to engage them with stories that you understand and can relate to their experience. When I think of the kind of story I want to tell to undergraduate students, I think of the *Vasa*, a ship and story that I think should be better known among engineers.

I first learned of the *Vasa* from a T-shirt at a conference in 1990. A company that a friend had started used the cross section of the ship to lampoon the ISO OSI effort on network protocols. "Another 7 Layer Model That Failed," read the caption. The connection was that ISO had seven layers and the *Vasa* had seven decks, but when I found out why the *Vasa* had tragically failed, I became fascinated, because it was such a classic engineering failure story.

The *Vasa* was built between 1626 and 1628 for King Gustavus Adolphus of Sweden, who was, at that time, attempting to rule the Baltic Sea. In the 17th century, rulers were expected to be capable of more than just giving orders, so Adolphus not only organized wars, he also helped design the ships of his naval fleet. At the time, Swedish warships had one deck of cannons on each side from which they fired fusillades at enemy ships, sometimes even hitting and damaging the other ships. When the *Vasa* was commissioned, this single row of cannons was considered state of the art.

Sometime during the construction of the ship Adolphus found out that the Poles had ships with two decks of guns, so he modified the design of the *Vasa* to have a second gun deck. This would have made it the most powerful naval vessel of the time, capable of delivering a broadside of devastating proportions. The men he had contracted to build his ships attempted to explain that the ship had too little ballast to support two gun decks and that the resulting ship would likely be unsafe to sail. The king insisted, just like, say, many project managers that his orders should be followed. On a software project you can quit, but if the king is your boss, you might lose more than your job; you might, say, lose your head, so the project went forward.

In 1628 the ship was finally ready for QA (quality assurance) testing. Seventeenth-century QA of ships was a bit different from what might happen today. Thirty sailors were selected and asked to run back and forth, port to starboard, across the deck of the ship. If the ship didn't tip over and sink, then the ship passed the test. You did not want to be on the QA team in 1628. After only three runs across the deck, the *Vasa* began to tilt wildly, and the test was canceled. The test may have been canceled, but not the project. This was the king's ship, after all, and she would sail. And sail she did.

On August 10, 1628, in a light breeze, the *Vasa* set sail. She was less than a mile from dock when a stiff breeze knocked her sideways. She took on water and sank in full view of thousands of onlookers. Approximately 30 to 50 sailors were killed when they were either trapped in the ship or unable to swim to shore.

In response to the catastrophe, the king wrote a letter insisting that incompetence had been the reason for the disaster. He was, of course, correct, but not in the way he might have envisioned. An inquest was held, and the surviving members of the crew, the captain, and the shipbuilders were questioned as to the state of the crew and the ship at the time of the incident. The mostly unstated belief by the end of the inquest was that the design had been a failure and the designer had not listened to the builders about the shortcomings of the design. Of course, the king could not be held at fault, so the final verdict was an "act of God." As a related aside, the disaster was also a huge economic loss for Sweden.

Now, this story may not be as well written as *Frankenstein*, but it's a much more direct warning about engineering failures. I think the funniest, or saddest, part of this story is how modern it is. Nothing has changed since 1628. People still fail to communicate, leading to failures of disastrous proportions. Egos get in the way, and mysterious supernatural forces are blamed for human failings. It's all kind of obvious in a really sad way.

In the 1960s the *Vasa* was raised from the bottom of the bay in which it had sunk and eventually placed in a museum in Stockholm. I visited the *Vasa* in 2000 as part of the SIGCOMM conference. The whole story is told there on plaques hanging on the walls. It's a museum all engineers ought to visit at least once.

KV

5.2 What Color Is Your...?

Why should I care what color the bikeshed is?

Poul-Henning Kamp

The habit of humans to argue over minor details rather than trying to understand large and complex systems has been well-documented, and will only get worse as the world, and the systems we put into the world, grow larger.[1] KV often blames this tendency on the marketing teams he's been exposed to as they seem to have nothing better to do than to argue over the more trivial details of a project. Unfortunately, technical people can also get stuck in the same types of weeds. Even though what's being argued about is different: marketing likes to pick nits about look and feel, and engineering likes to pick nits about coding style, but it's the fact that the arguments are over nearly unimportant details that is the root of the problem.

People will find any reason to push a personal agenda; sometimes they seem to get off on just feeling as if they were right and others were wrong. The need to pick the unimportant nits in a system is a form of behavior that can quickly become poisonous and turns into a form of bullying whereby the nitpicker gets their way even if that's not the right way, simply because no one has the energy to fight with them. Being able to call out when an argument goes down a rabbit hole is a good skill to learn, and having a saying, or watch cry, that everyone recognizes to indicate that that's where the conversation has gotten to can be critically important to quashing unproductive arguments.

On the FreeBSD Project the watch cry for this behavior, for many years, has been "Bike Shed" for reasons that will now become clearer.

---

1. C. Northcote Parkinson: Parkinson's law: or, the pursuit of progress / C. Northcote Parkinson; with illustrations by Osbert Lancaster, English, 1957, p. 1 v.

Dear KV,

Last week one of our newer engineers checked in a short program to help with debugging problems in the code that we're developing. Even though this was a test program, several people read the code and then commented on the changes they wanted to see. The code didn't have any major problems, but it seemed to generate a lot of e-mail for what was being checked in. Eventually the comments in the thread were longer than the program itself. At some point in the thread the programmer who submitted the code said, "Look, I've checked in the code; you can paint the bike shed any color you want now," and then refused to make any more changes to the code. I can understand his frustration at nitpicking comments, but what did he mean about painting the bike shed?

Maybe I'd Like Green

Dear Green,

A bike shed is a shed that you park your bike in, and since all such sheds need to be painted to protect them from the weather, they must be painted in some color. Color is very important to some people. OK, that's not what you're really asking.

What you witnessed was, unfortunately, a typical reaction to simple changes in a code base. Have you ever noticed that when someone checks in some complex and, oftentimes, horrific piece of code, the check-in is greeted with an almost deafening silence? That is often because the people who should be reviewing such code just don't have the time. Unfortunately, lots of people do have the time to review a 10- or 50-line change, and since they feel guilty for not having checked the bigger pieces of code, they nitpick the small pieces.

The explanation for why this occurs was first given by C. Northcote Parkinson, who wrote a book about management (*Parkinson's Law and Other Studies in Administration*, Ballantine Books, 1969). He stated that if you were building something complex, then few people would argue with you because few people could understand what you were doing. If you were building something simple, say, a bike shed, which most anyone could build, then everyone would have an opinion. Unfortunately, as you learned, it's not enough just to have an opinion, but most people feel they should express that opinion.

The engineer who wrote the original code clearly figured out that he was trapped in a pointless loop and decided to break out of it by telling people to paint the shed any color they liked. He or she was probably pretty sure that no one would actually change the color of the shed, but by not engaging in the pointless loop was able to let people get back to ignoring more important parts of the code, which is what they had been doing previously.

KV

P.S. For another version of bike-shed story go to:

```
http://www.freebsd.org/doc/en_US.ISO8859-1/books/faq/
misc.html#BIKESHED-PAINTING
```

P.P.S. I like my sheds to be fushcia.

5.3 Broken Builds

> *I don't care if it works on your machine! We are not shipping your machine!*
>
> Vidiu Platon

Most people will not believe this, but the majority of koders are optimists. I exclude from this set those of us who work in and around computer security, but your garden-variety koder (yes, they are organically grown, even if not locally sourced) is, nine times out of ten, an optimist. Given what we do, you can see how this has to be the case. Getting a large system, or even just a difficult piece of code, to work is such a fantastically annoying and difficult task that by the time the code builds most koders are then sure that it also works, leading to the oft-repeated joke, "It builds; now ship it." That little in-joke would be funnier if fewer people actually acted on it.

There are downsides to this optimism, most of which are called out in various letters throughout this book, but the one that is most pertinent here is a lack of testing before committing code. Building software is, for 90 percent or more of the world a shared endeavor. Not testing code before committing it is one of the cardinal sins of software development. What is most frustrating about this continued tendency among koders is that computers are now so fast that only the largest systems take more than a few minutes to run through a test build. Build infrastructure is not only available at large corporate workplaces but is now available to even the smallest open source project, and in fact for many systems the build and test infrastructure could run on a laptop.

The other complicating factor for koders is that often we do not consider people besides ourselves when we are working. The single-minded focus that is required to bend some nasty piece of code to our will forces us to remove consideration of all things outside of the kode right in front of us. That lack of consideration for others is another reason that people will test their changes in solitude, commit them, and then walk away to get another beverage. KV is not suggesting that we remain trapped in a develop, merge, test, commit cycle where we spend all our time finding out that everyone else in the project screwed up the shared repository of work, but that that practice needs to become common for everyone on a project.

Breaking shared infrastructure should be frowned on, just like pissing in a public pool.

Dear KV,

Is there anything more aggravating to programmers than fellow team members checking in code that breaks a build? I find myself constantly tracking down minor mistakes in other people's code simply because they didn't check that their changes didn't break the build. The worst part is when someone has broken the build and they get indignant about my pointing it out. Are there any better ways to protect against these types of problems?

Made to be Broken

Dear Made,

I know you, and everyone else, are expecting me simply to rant about how you should cut off the tips of the pinkies of the offending parties as a lesson to them and a warning to others about carelessness. While that might be satisfying, it's illegal in most places and, I'm told, morally wrong.

A frequently broken build is a symptom of a disease, but it is not the disease itself. It indicates problems in any of the following three areas: management, infrastructure, or software architecture.

Management is the area that most quickly comes to mind when there is a team- or project-wide problem. The belief of most of the workers on a project tasked with writing and verifying code and systems is that project-wide problems need to be solved by Mommy (aka the project lead or the manager). Unfortunately, Mommy can remind people only so often to clean up their rooms, to tie their shoes, and not to check in broken code.

One of the best solutions to the problem of people not checking their code before they check it in is peer pressure. Anyone who checks in code without compiling it first ought to feel embarrassed by such a mistake, and if not, the other people around them should strongly encourage them to feel embarrassed. Shame, it turns out, is a strong motivator for avoiding antisocial behavior. Like many or perhaps all of KV's suggestions, shaming can be taken too far, but I suggest you try it and see how it works.

Depending on Mommy to tell off the misbehaving kids becomes tiresome both for you and for the project manager after a while. What you want to see is a good working culture develop, one in which people know that breaking the build is like taking a dump in the middle of the break room; funny once, but usually unacceptable.

Poor infrastructure can also lead to suffering with frequently broken builds. One thing that continues to amaze me is how computer hardware gets cheaper, and yet companies continue to coast along without a nightly, or more frequent, build system. For the price of a single desktop computer and a few days of scripting, most teams can have a system that periodically updates a test build of their code, builds it, and sends e-mail to the team if the build fails. The amount of time saved by such a system is easily measurable.

Subtract 1 from the number of programmers on a team. Multiply the resulting number by the number of hours it usually takes to figure out who broke the build, find them, shame them, and have them fix the build. Now multiply *that* number by the average hourly wage of each person on the team, and you have a rough idea of how much time and money was wasted by not having periodic builds. We won't get into periodic testing, which can save even more time and money, because if your build is always broken, you clearly have not achieved a sufficient level of sophistication to move on to nightly tests.

Even though the broken code will still get into the system, with a periodic build system the offending person will find out fairly quickly that he or she broke the build and hopefully will admit it in an e-mail ("I broke the build, hang on a second") and then repair the error. While this is still suboptimal, it is far better than what you had before.

Sometimes it is the build system itself that is the source of the problem. Many modern build systems depend heavily on caching derived objects, as well as the parallelization of the build process. While a parallel build process can provide you results more quickly, it can often lead to build failures that are false positives. Trying to build an object that requires another object to be created first, such as an automatically created include file, always leads to trouble. Maintaining the list of dependencies by hand is an error-prone, but often necessary, process. If you are using a build system that depends on caching and uses parallel builds, then your problems may lie here.

Now we come to the final area that is the cause of build problems. The way in which a piece of software is put together, frequently referred to as its architecture, often impacts not only how the software performs when it runs, but also how it is built. I hesitate to use the word "architecture" since overuse of the term has led to the unfortunate proliferation of the job title software architect, which is far too often a misnomer.

If all the components of a software system are too interdependent, then a change to one can result in an injury to all. A lack of sufficient modularization is often a problem when software ships, but it is definitely a problem when the software is being compiled. When a change to an include file in one area leads to the build breaking in another area, then your software is probably too heavily interlinked, and the team should look at breaking the pieces apart. Often such links come from careless reuse of some part of the system. Careless reuse is when you look at a large abstraction and think, "Oh, I really want this version of method X," where X is a small part of the overall abstraction, and then you wind up making your code depend not just on the small part you want, but on all of the parts that X is associated with. If you get to the point where you know that it's neither carelessness nor poor infrastructure that is leading to frequent build failures, then it's time to look at the software architecture.

Now you know the three most basic ways to alleviate frequent build breakage: shaming your teammates, adding some basic infrastructure, and finally improving the software architecture. That ought to keep you out of jail, for now.

KV

5.4 What Is Intelligence?

> *A robot may not injure a human being or, through inaction, allow a human being to come to harm.*
>
> First of the Three Laws of Robotics from Isaac Asimov

In the late teen years of the 21st century it has been difficult to avoid three types of over-hyped technologies: Internet of Things, block-chain, and AI. Thus far, KV has been able to avoid talking about block-chain, but has certainly enjoyed the various cycles of schadenfreude around it. These pages already include discussions that touch upon the Internet of Terror, or as another colleague has pointed out, "The S in IoT is for Security!" When it comes to artificial intelligence, there has been only one letter worth replying to on the topic so far.

Dear KV,

Our company is looking at handing much of our analytics to a company that claims to use "Soft AI" to get answers to questions about the data we have collected via our online sales system. I've been asked by management to evaluate this solution, and throughout the evaluation all I can see is that this company has put a slick interface on top of a pretty standard set of analytical models. I think what they really mean to say is "Weak AI" and that they're using the term Soft so they can trademark it. What is the real difference between soft (or weak) AI and AI in general?

Feeling Artificially Dumb

Dear AD,

The topic of AI hits the news about every 10 to 20 years, whenever a new level of computing performance becomes so broadly deployed as to enable some new type of application. In the 1980s it was all about expert systems. Now we see advances in remote control (such as military drones) and statistical number crunching (search engines, voice menus, and the like).

The idea of artificial intelligence is no longer new, and, in fact, the thought that we would like to meet and interact with nonhumans has existed in fiction for hundreds of years. Ideas about AI that have come out of the 20th century have some well-known sources, including the writings of Alan Turing and Isaac Asimov. Turing's scientific work generated the now famous Turing test, by which a machine's intelligence would be judged against a human one; and Asimov's fiction gave us the Three Laws of Robotics, ethical rules that were to be coded into the lowest-level software of robotic brains. The effects of the latter on modern culture, both technological and popular, is easy to gauge, since newspapers still discuss advances in computing with respect to the three laws. The Turing test is, of course, known to anyone involved in computing, perhaps better known than the halting problem (https://en.wikipedia.org/wiki/Halting_problem), much to the chagrin of those of us who deal with people wanting to write "compiler-checking compilers."

The problem inherent in almost all nonspecialist work in AI is that humans actually don't understand intelligence very well in the first place. Now, computer scientists often think they understand intelligence because they have so often been the "smart" kid, but that's got very little to do with understanding what intelligence actually is. In the absence of a clear understanding of how the human brain generates and evaluates ideas, which may or may not be a good basis for the concept of intelligence, we have introduced numerous proxies for intelligence, the first of which is game-playing behavior.

One of the early challenges in AI (and for the moment I'm talking about AI in the large, not soft or weak or any other marketing buzzword) was to get a computer to play chess. Now, why would a bunch of computer scientists want to get a computer to play chess? Chess, like any other game, has a set of rules, and rules can be written in code. Chess is more complicated than many games, such as tic-tac-toe (a game that is used to demonstrate to another fictional computer in the 1983 film *War Games* that nuclear war is unwinnable), and has a large enough set of potential moves that it is interesting from the standpoint of programming a winning set of moves or a strategy. When computer programs were first matched against human players in the late 1960s, the machines that were used were, by any modern concept, primitive and incapable of storing a large number of moves or strategies. It wasn't until 1996 that a computer, the specially built Deep Blue, beat a human Grandmaster at the game.

Since that time, hardware has continued its inexorable march toward larger memories, higher clock speeds, and now, more cores. It is now possible for a handheld computer, such as a cellphone, to beat a chess Grandmaster. We have had nearly 50 years of human/computer competition in the game of chess, but does this mean that any of those computers are intelligent? No, it does not, for two reasons. The first is that chess is not a test of intelligence; it is the test of a particular skill or the skill of playing chess. If I could beat a Grandmaster at chess and yet not be able to hand you the salt at the table when asked, would I be intelligent? The second reason is that considering chess to be a test of intelligence was based on a false cultural premise that brilliant chess players were brilliant minds, more gifted than those around them. Yes, many intelligent people excel at chess, but chess, or any other single skill, does not denote intelligence.

Shifting to our modern concepts of soft and hard AI or weak and strong, or narrow and general, etc., we are now simply reaping the benefits of 50 years of advancements in electronics, along with a small set of improvements in applying statistics to very large data sets. In fact, improvement in the tools that people think are AI is, in no small part, a result of the vast amount of data that it is now possible to store.

Papers on AI topics in the 1980s often postulated what "might be possible" once megabytes of storage were commonly available. The narrow AI systems we interact with today, such as Siri and other voice-recognition systems, are not intelligent; they cannot pass the salt, but they can pick out features in human voices and then use a search system, also based on stats run on large data sets, to somewhat simulate what happens when we ask another person a question. "Hey, what's that song that's playing?" Recognizing the words is done by running a lot of stats on acoustic models and then running another algorithm to throw away the superfluous words ("Hey," "that," "that's") to get "What song is playing?" This is not intelligence, but, as Arthur C. Clarke famously quipped, "Any sufficiently advanced science is indistinguishable from magic."

All of which is to say that KV is not surprised in the least that when you peek under the hood of "Soft AI," you find a system of statistics run on large data sets. Intelligence, artificial or otherwise, remains firmly the domain of philosophers and, perhaps, psychologists. As computer scientists, we may have pretensions about the nature of intelligence, but any astute observer can see that there is a lot more work to do before we can have a robot pass us the salt, or tell us why we might or might not want to put it on our slugs before eating them for breakfast.

KV

5.5 Review the Design

> *Perfection is achieved, not when there is nothing more to add, but when there is nothing left to take away.*
>
> Atoine de Saint Exupery

Design reviews are an excellent way to address the larger issues in both a software system and a human system. Since it is still the case that humans design and implement the systems we are all subjected to, a design review is a good start to understanding a large piece of software; it is also a good way to depersonalize any problems found in the design.

A design review can happen at nearly any point in a project, but they're usually best held once there is a design to review. Fairly often systems are implemented without a formal design, in which case the design review is often a post-mortem run against one system, to inform the design of the follow-on system that will need to replace the nondesigned, organically grown mess that you find yourself in. No matter at what point the design review happens, it should take the same structure, as described in the following letter and response.

One key to a good design review not brought out in the original response coming up is that it must always be impersonal. The review is of the design, not the person who came up with the design nor the people who implemented the system. Too often people use in-person review meetings to attack the people and not the design, which is both wrong and counterproductive. A clear way to know the difference between a personal and impersonal review is to watch for phrases like, "What made you think *x*?" and "Why did you implement *y*?" which are clearly personal and accusatory. KV knows these are personal and accusatory because he uses "What made you think *x*?" on a daily basis—but not in design reviews, mostly in casual conversation. Phrases that show an interest in the design itself are more of the form of, "What data do you have to show that this is a more efficient coupling between these two components?" and "What is the plan to handle extensions to this system?"

Design reviews go better when there is a written design to review. It is depressing how often KV has to ask, "Did you write this down? If so, where?" Over the years I've been able to mostly keep the snark out of my voice when asking this question. The nice thing about a design review is that if the design has not been written down, the notes from the review should make an excellent start of a design document, which means that even if you went into the review without a document, you damned well better come of out of it with one. KV likes to be the one taking the notes, because the person who takes the notes is the one who controls the history.

Dear KV,

I was recently hired as a mid-level web developer working on version 2 of a highly successful but outdated web application. It will be implemented with ASP.NET WebAPI. Our architect designed a layered architecture, roughly like Web Service / Data Service / Data Access. He noted that data service should be agnostic to Entity Framework ORM (object-relational mapping), and it should use unit-of-work and repository patterns. I guess my problem sort of started there.

Our lead developer has created a solution to implement the architecture, but the implementation does not apply the unit-of-work and repository patterns correctly. Worse, the code is really hard to understand, and it does not actually fit the architecture. So I see a lot of red flags coming up with this implementation. It took me almost an entire weekend to work through the code, and there are still gaps in my understanding.

This week our first sprint starts, and I feel a responsibility to speak up and try to address this issue. I know that I will face a lot of resistance, just based on the fact that the lead developer wrote that code and understands it more than the alternatives. He may not see the issue that I will try to convey. I need to convince him and the rest of the team that the code needs to be refactored or reworked. I feel apprehensive, because I am like the new kid on the block trying to change the game. I also don't want to be perceived as Mr. Know-It-All, even though I might be a little more opinionated than I should be sometimes.

My question is, how can I convince the team that there is a real problem with the implementation without offending anyone?

Opinionated

Dear Opinionated,

Let me work backward through your letter from the end. You are asking me, Kode Vicious, how to point out problems without offending anyone? Have you read any of my previous columns? Let's just start out with the KV ground rules: it's only the law and other deleterious side effects that keep me on the "right" side of violence in some meetings. I'd like to think a jury of my peers would acquit me should I eventually cross to the wrong side, but I don't want to stake my freedom on that. I will try my best to give you solutions that do not land you in jail, but I will not guarantee them not to offend.

Trying to correct someone who has just done a lot of work, even if, ultimately, that work is not the right work, is a daunting task. The person in question no doubt believes that he has worked very hard to produce something of value to the rest of the team, and walking in and spitting on it, literally or metaphorically, probably crosses your "offense" line, at least I think it does. I'm a bit surprised, since this is the first sprint, that there is already so much code written. Shouldn't the software have shown up after the sprints established what was needed, who the stakeholders were, etc.? Or was this a piece of previously existing code that was being brought in to solve a new problem? It probably doesn't matter, because the crux of your letter is the fact that you and your team do not sufficiently understand the software in question to be comfortable fielding it.

In order to become more comfortable with the system, there are two things to call for: a design review and a code review. These are not actually the same things, and KV has already covered how to conduct a code review ["Kode Reviews 101." Communications of the ACM 52(10): 28–29. (October 2009)]. Let's talk now about a design review.

A software design review is intended to answer a basic set of questions:

1. How does the design take inputs and turn them into outputs?
2. What are the major components that make up the system?
3. How do the components work together to achieve the goals set out by the design?

That all sounds simple, but the devil is in the level of the details. Many software developers and systems architects would prefer that everyone but themselves see the systems they have built as black boxes, where data goes in and other data comes out, no questions asked. You clearly do not have the necessary level of trust with the software you're working with to allow the lead developer to get away with that, so you should call for a design review where you take the lid off the box and poke around at the parts inside. In fact, questions 2 and 3 are going to be your main tools for figuring out what the software does and whether or not it is suitable for the task.

When I have to interview people for jobs, I always ask them questions about systems they have worked on while we draw out the block diagram on a whiteboard: What are the major components? How does component A talk to component B? What happens if C fails? I'm trying to transfer their mental images of their software into my own mind, of course without either going mad or having a nasty flashback. Some pieces of software are best left outside your mind, but hopefully that's not going to be the case with the system you're working with.

Remember that every box that this person draws can be opened if you think you're not getting sufficient detail. Much like the ancient game show, *Let's Make a Deal*, it is always OK for you to ask, "What's behind door number 1, Monty?" Of course, you might find that it's a goat, but hopefully you find that it's a working set of components that are understandable to you and the team.

The one thing not to do in a design review is turn it into a code review. You are definitely not interested in the internals of any of the algorithms, at least not yet. The only code you might want to look at are the APIs that glue the components together, but even these are best left abstract, so that the amount of detail does not overwhelm you. Remember that the goal is always to get the big picture rather than the fine details, at least in a design review.

Coming back to the question of offense, I have found only one legal way to avoid giving offense, and that is always to phrase things as questions. Often called the Socratic method, this can be a good way to get people to explain to you, and often to themselves, what they think they are doing. The Socratic method can be applied in an annoyingly pedantic way, but since you're trying not to give offense, I suggest that you play by a few useful rules. First, do not hammer the person with a relentless list of questions right off. Remember that you are trying to explore the design space in a collaborative way; this is not an interrogation. Second, leave spaces for the people you're working with to think. A pause doesn't mean they don't know; in fact, it might be that they're trying to adjust their mental model of the system in a way that will be beneficial to everyone when the review is done. Lastly, try to vary the questions you ask and the words you use. No one wants to be subjected to a lot of, "And then what happens?"

Finally, I find that when I'm in a design review and about to do something that might give offense, such as throwing a chair or a whiteboard marker, I try to do something less obvious. My personal style is to take off my glasses, put them on the table and speak in a very calm voice. That usually doesn't offend, but it does get people's attention, which leads them to concentrate harder on working to understand the problem we're all trying to solve.

KV

5.6 The Naming of Hosts

> *Wait, what did you name your laptop?*
>
> Several people on seeing KV's screensaver

One of the main categories of fights that technical folks have most often, *outside* of a piece of code, is naming schemes. KV has personally used some possibly questionable naming schemes in the past, for instance the time I named a group of workstations after ex-boyfriends, or the series of systems I owned that were all pejoratives. Both of these schemes had the very positive side effect of being easy to remember, for me at least. Everyone has their favorite, but what's the best way to choose one? Is there a best way? Well, KV took a crack at some methods for the naming of hosts.

Dear KV,

An argument recently broke out between two factions of our systems administration team concerning the naming of our next set of hosts. One faction wants to name machines after services, with each host having a numeric suffix, and the other wants to continue our current scheme of each host having a unique name, without a numeric string. We now have so many hosts that any unique name is getting quite long and is annoying to type. A compromise was recently suggested whereby each host could have two names in our internal DNS (Domain Name System), but this seems overly complicated. How do you decide on a host-naming scheme?

Anonymous

Dear Anonymous,

I refer you to T.S. Eliot, who pointed out sort of:

> The Naming of Hosts is a difficult matter,
> It isn't just one of your holiday games;
> You may think at first I'm as mad as a hatter
> When I tell you, a host must have *three different names.*

"The Naming of Cats" (not Hosts) is a poem in T. S. Eliot's poetry book, *Old Possum's Book of Practical Cats*, and its stage adaptation is Andrew Lloyd Webber's popular musical *Cats*. The poem describes to humans how cats get their names. I took some liberties with Eliot's wording as others have done before me and extended the analogy to describe the naming of hosts. But given that T. S. Eliot died just about the time the first minicomputers were being designed, I don't think he had host names in mind when he wrote his poem. And that's a good thing, because if you think two names are bad, three would only be worse!

The naming of hosts is a difficult matter that ranks with coding style, editor choice, and language preference in the pantheon of things computer people fight about that don't matter to anyone else in the whole world. What's even more annoying or amusing, but actually annoying, is that if you're in the wrong bar at the wrong time, you'll have to hear drunken systems administrators fighting about naming schemes and crying in their beers over the names they lovingly gave to hosts at their previous companies. What a way to ruin a good bender!

Giving something a name has a simple purpose: to make it understandable and memorable to a community of people. Naming your variables foo, bar, and baz is amusing in a short example program, but you wouldn't want to maintain 100 lines of code written like that. The same is true of host names. Hosts have names because people need to know how to get to them either to use their services or to maintain them, or both. If people weren't involved, hosts could simply be identified by their Internet addresses. Unfortunately, host naming is an instance where geeks like to get creative. Even more unfortunately, geeks don't always know the difference between creative and annoying. It's all very well to decide that your hosts should be named after Star Trek, Star Wars, or Tolkien or Twilight characters. With Tolkien you can probably write (and, dear God, someone has probably already done so) a script to generate new names based on his works, just in case The Hobbit, The Lord of the Rings trilogy, and The Silmarillion didn't have enough ridiculous names in them to begin with!

Everyone has a naming horror story. My first was at a university where the hosts were named after rivers. That would have been fine if you could remember how to spell Seine, but once you run out of nice short names, you get to Mississippi and Dnjeper. That's what I want to do when I remotely log in to a host, I want to think in my head, "M-I crooked letter crooked letter I crooked letter crooked letter I hump back hump back I," which is how I and many other American schoolchildren learned to spell Mississippi. I could go on and on about this, but then I would sound like those folks I mentioned who were ruining my bender. Here, therefore, is a short guide to picking host names.

A name that you're going to use on a daily basis needs to be easy to type. That means no silent letters, such as in Dnjeper, and nothing that's too long, like thisisthehostthatjackbuilt.

It's a good idea to choose names that everyone you work with can pronounce. With globalization, finding pronounceable names has become more difficult, since some people can't pick up "L" vs. "R," or understand whether you just used a double "o" or a single "o," and diphthongs will kill you (no, diphthongs are not a new Brazilian bathing suit). The main point here is to avoid picking a name with a lot of sounds that are difficult to translate into typing. Typing is still faster than using a voice recognition system; so remember, these names will have to be typed.

If you're going to use services as names, make sure you can replace the systems behind the names without hiccups. It should be obvious that everyone is going to be annoyed if they have to use mail2.yourdomain.com when mail.yourdomain.com goes down. (This point isn't really about naming, because any sysadmin worth his or her paycheck can build a system like this; but I've seen it done the wrong way, so I wanted to state it for the record.)

Avoid at all costs having two different, unrelated names for the same thing. In fact, this is true in code and host names. If you have two similar services and you want two different names, make it completely obvious how to map one name to the other and back. It is maddening to have the kind of back and forth where one person asks,

> "Hey, can I reboot fibble?"

> "Yes."

> And then someone asks,

"Who rebooted mail1?"

"But I didn't know it was mail1; I thought it was fibble."

Finally, try to avoid being cute. I know that giving this piece of advice is basically tilting at windmills, but I have to say that people who name their mail servers male and female make my normally icy blood boil.

KV

5.7 Hosting an Interview

*Can you draw me a block
diagram of the system?*

First serious question in any
interview

What's the best way to figure out if the human in front of you is someone you want touching your code? There have been many books written on interviews and interviewing, but KV has some simple ideas so that you, as an interviewer, can, in the short period of time that HR allotted you, figure out if that lump of flesh across the conference room table can actually think, inside or outside a box.

Dear KV,

My work group has just been given approval to hire four new programmers, and now all of us have to interview people, both on the phone and in person. I hate interviewing people. I never know what to ask. I've also noticed that people tend to be careless with the truth when writing their resumes. We're considering a programming test for our next round of interviewees, because we realized that some previous candidates clearly couldn't program their way out of a paper bag. There have to be tricks to speeding up hiring without compromising whom we hire.

Tired of Talking and Not Coding

Dear Tired of Talking,

My preferred model for judging prospective job candidates is the one used by the U.S. Marines: beat the candidate down until he or she has no remaining shreds of individuality or self-respect, and if the candidate still wants to work for you, hire that person because at that point you own his or her soul. Unfortunately, every time I suggest "bootcamp hiring" at the office, our legal department has a conniption fit, and so I have yet to implement and test this method. In the absence of being able to judge people's abilities under fire, you have to use less direct, subtler approaches.

The true goal of any interview is for both parties (the interviewer and interviewee) to work out whether the person can do the work and is a good fit with the rest of the group. There are many brilliant programmers out there whom I would never hire because the detrimental impact of their character defects on the rest of the team would outweigh their abilities as coders. The question to ask is, "How do I determine, in 30 to 60 minutes, if this person can do the work we need, and in a way that I can put up with him or her for 10 hours a day, five days a week, and possibly for years on end?" That's a lot to ask of such a short meeting.

Figuring out if someone has the knowledge you think is needed for a job is probably the easiest component of an interview. You don't really need to give the candidate a programming test at this point; all you need to do is to ask questions that you've recently answered at work. This assumes that the person will be doing the same kind of work you've been doing, and that's usually a safe assumption (programmers are, happily, rarely asked to interview people for the accounting department). I tend to start with basic questions first, and you should not feel that a senior person is above answering simple questions. Just because someone has a lot of experience does not excuse him or her from knowing how to work with, for example, linked lists. If a person has a lot of experience, then he or she will exhaust your trivial questions pretty quickly and you can move on to more difficult problems.

Once you've established that the person has the basic knowledge for the job, you need to figure out, at least at a high level, how he or she solves problems. At this point I find a whiteboard to be the best tool for the interview. I always make candidates for programming jobs describe, in block diagram form, a system they're familiar with. If a programmer or software engineer cannot describe a system as a block diagram, I am pretty unlikely to hire that person. A candidate might be brilliant and might understand the system he or she worked on, but a candidate who cannot explain it to another person will be useless in any work group.

After a candidate has described a system to my satisfaction, however, I always ask the following question: "If you had had more time, what would you have changed or what feature would you have added?" This type of open-ended question is extremely important. Someone who cannot answer this question is pretty much an automaton that simply implements the will of others. I don't like working with automatons; I like working with thinking human beings who have opinions on what they're building and who are always thinking about how they can extend the systems they're working on. Good programmers understand that their systems are never really complete and that there is always something else they could have done, if only they'd had more time.

While some companies use a programming test to evaluate candidates, I prefer two different ways of achieving the same thing. Many years ago I worked for a company that asked you to submit a piece of code it could compile and run. The code needed to be well-documented and simple enough for someone to figure out in less than an hour. I much prefer real examples of code to asking someone to code a bubble sort on paper. With the advent of widespread participation in open source projects, this kind of test is less necessary, because you can often just search for a programmer's name and see some code, somewhere, that the programmer has added to an open source project. However you do it, make sure to get a sample of the code, because that will tell you far more about the programmer than how he or she codes on a legal pad. If you're the paranoid type and you don't trust the person not to send a friend's code, then ask about the code during the interview. If the person is bluffing, then you'll know it pretty quickly; and if the person did submit a friend's code but can bluff his or her way through explaining it, then you should hire that candidate anyway.

The other way that I like to quiz programmers on code is to provide a piece of code that's broken in some way and ask them to find the bugs. Much of a programmer's workday is spent debugging, and I value that ability almost as highly as I value the ability to write bug-free code. Even if the programmers themselves write bug-free code, a rather highly unlikely occurrence, they're going to work on other people's code, and it's important that they have the ability to quickly analyze code that's not their own.

Some interviewers like to give brainteasers, but I don't find these to be useful, unless they relate to coding in the real world. Brainteasers fail on several levels. On one level, a person can simply memorize most of the popular ones. A quick search shows that there are more than 2,000 books on how to ace interview questions, with some books specifically about software. If candidates find out that your company is the type of company that uses brainteasers, then they'll just brush up before the interview and quite possibly get by your interviewers.

Another reason to avoid the brainteaser approach is that there are some very good programmers who are terrible at brainteasers, so you'll miss out on some good people because they didn't seem smart on your tests. Let's face it, most programmers do not spend their days looking at coding problems and trying to relate them to brainteasers. They look at problems and try to work out in code how to solve them. You're hiring programmers, not game-show contestants.

I realize that you originally asked for a way to speed up the hiring and interview process. I've already made one suggestion that can speed things up: have the interviewee send you code before coming to the interview. I suggest you do this after the phone screen but before the in-person interview, because if the person sends you code that's complete crap, you can forgo the interview.

The second thing to do to avoid wasting people's time is to evaluate, after each of your team members talks to the person, whether to send that candidate on to the next team member. While it might be embarrassing to send someone home after the first one or two interviewers have talked to the candidate, it's a hell of a lot less stressful than having the whole team go through the motions of interviewing someone, just to be nice. Just as in code, it's better to fail early.

KV

5.8 Mythical

Nine mothers cannot make a baby in one month.

Frederick P. Brooks

Dr. Fred Brook's classic text on software design and development *The Mythical Man Month* appears twice in the columns of Kode Vicious. The first time resulted in Dr. Brooks very kindly offering to replace the copy I'd read at university and sold to buy 4th of July beer with a copy that he'd signed himself. To say that I was shocked to know that he read KV is to put it mildly, but I was very happy to have a new copy for myself, and that one will not be sold for beer or any other inebriates. The longer of the two pieces referencing this classic work appears below, in which KV tries to answer the question, "How many prototypes should you expect to throw away?"

Dear KV,

In his book *The Mythical Man-Month*, Frederick P. Brooks admonishes us with grandfatherly patience to plan to build a prototype and to throw it away. You will anyway.

At one point this resulted in a fad-of-the-year called prototyping (the programming methodology formerly known as trial and error), demonstrating that too little and too much are equally as bad.

What is your view of creating prototypes, and, particularly, how faithful does a prototype need to be to resolve the really tricky details, as opposed to just enabling the marketing department to get screen shots so they can strut the stuff?

Signed, An (A)typical Engineer

Dear Atypical,

What do you mean by "formerly known as trial and error"!?! Are you telling me that this fad has died? As far as I can tell, it's alive and well, though perhaps many of its practitioners don't actually know their intellectual parentage. Actually, I suspect most of its practitioners can't spell intellectual parentage.

Alas, it is often the case that a piece of good advice is taken too far and becomes, for a time, a mantra. Anything repeated often enough seems to become truth. Mr. Brooks's advice, as I'm sure you know, was meant to overcome the "it must be perfect" mantra that is all too prevalent in computer science. The idea that you can know everything in the design stage is a fallacy that I think started with the mathematicians, who were the world's first programmers. If you spend your days looking at symbols on paper, and then only occasionally have to build those symbols into working systems, you rarely come to appreciate what happens when the beauty of your system meets the ugly reality that is hardware.

From that starting point, it's easy to see how programmers of the 1950s and 1960s would want to write everything down first. The problem is that a piece of paper is a very poor substitute for a computer. Paper doesn't have odd delays introduced by the speed of electrons in copper, the length of wires, or the speed of the drum (now disk, soon to be flash). Thus, it made perfect sense at the time to admonish people just to build the damned thing, no matter what it was, and then to take the lessons learned from the prototype and integrate them into the real system.

The increasing speeds of computers since that advice was first given have allowed people to build bigger, faster, and certainly more prototypes in the same amount of time that they could have built a single system in the past. The sufferers of prototypitis are really just chicken. Not putting a line in the sand is a sign of cowardice on the part of the engineer or team. "This is just a prototype" is too often used as an excuse to avoid looking at the hard problems in a system's design. In a way, such prototyping has become the exact opposite of what Mr. Brooks was trying to do. The point of a prototype is to find out where the hard problems are, and once they are identified, to make it possible to finish the whole system. It is not to give the marketing department something pretty to show potential customers—that's what paper napkins and lots of whiskey are for.

Where do I stand on prototypes? The same place that I stand on layering or the breaking down of systems into smaller and smaller objects. You should build only as many prototypes as are necessary to find and solve the hard problems that result from whatever you're trying to build. Anything else is just navel-gazing. Now, don't get me wrong, I like navel-gazing as much as the next guy (perhaps more), but what I do when I delve into my psychedelia collection has nothing, I assure you, to do with writing software.

KV

5.9 The Obsolete Koder

> *Perhaps it is this specter that most haunts working men and women: the planned obsolescence of people that is of a piece with the planned obsolescence of the things they make. Or sell.*
>
> Studs Terkel

Remaining relevant in one's field is an important question and one that is rarely discussed in undergraduate or graduate education. Once a programmer starts working, they might pick up new skills more by accident than by design. How do we remain relevant in our work? Remaining relevant is the work of a lifetime, and if you do it well, you'll never be done.

Dear KV,

What is the biggest threat to systems administrators? Not the technical threat (security, outages, etc.), but the biggest threat to systems administrators as a profession?

A Budding Sysadmin

Dear Budding,

Career questions are quite a bit more difficult than technical questions because they require me to look into the future, and much as I might enjoy doing that, well, my doctor keeps telling me to lay off the hard drugs, at least during working hours.

I think the question you're really asking is, "What might make me obsolete?" and that is a question that anyone in any field, but particularly in a fast-moving technical field, should ask. The biggest risks to a systems administrator, then, are overspecialization and allowing others to define your job too narrowly; and proving your worth.

When most people think of overspecialization, they think of factory workers, who were, of course, made to specialize so that they could be better cogs in whatever means of production they were working on. The assembly-line worker who did one job for 10 years would have to be retrained when the machine that he or she worked with was changed, or, as was more likely, he or she was laid off, and a cheaper, younger worker was brought in as a replacement. Believing oneself immune to these types of problems because of current income or current perceived social class could be a career-ending mistake.

Overspecialization is a risk to anyone in a fast-moving field, in which some highly valued skill might be automated next week. I could even argue that the more valuable the skill, the more likely it is to be automated, because your corporate masters are interested in reducing their overhead so they can make points with their boss and get a bigger bonus. I have always felt that it's a good idea to have a broad set of interests in your area and then to have more than one area in which you can specialize. That way, if your particular specialty is suddenly made obsolete, then you have something else, which is interesting to you, with which to pay your bills.

How can you tell if you're overspecialized? The best indication is if your job is to repeat, over and over, the same task, and that task is designed and dictated by someone else. If your job is to configure systems, but does not include deciding how they're configured, then you're definitely at risk. At some point the configuration part, the repetitive process, will be automated, and if you haven't graduated to configuration architect, then you're likely to find yourself looking for new work. The issue isn't necessarily related to level, but it is related to scope. If you don't have sufficient scope to be making decisions, then you are simply a tool that is used by others, and tools get replaced.

Avoiding overspecialization is not difficult, but it takes work on your part. Taking a broad interest in your entire discipline helps, as does reading books and attending conferences and tutorials. The key is to choose your venues carefully so that you get as much exposure to areas with which you are unfamiliar. Looking at a list of books, conference sessions, or courses and picking the one that I know the least about is my favorite tactic. If you find yourself saying, "I have no use for X," then you better make damned sure you know that subject well and do not dismiss it out of hand.

The flip side of overspecialization is when someone else is defining your role for you. All businesses, and in particular large businesses, want to place their workers into well-defined boxes so that they can more easily calculate wages and benefits. The people who draw these boxes rarely understand what systems administrators are or what they do.

The usual way in whch these boxes are drawn is that the person doing the drawing does a web search for some terms, many of which are already woefully out of date, and then draws a box and puts your name in it. If you complain about this kind of treatment and you're lucky, they may even ask you to define your role, thereby doing their work for them. Let me recommend against defining your role as, "The god who makes it possible for you to get your work done." No matter how true this might be, no one likes it when you say that sort of thing. At this point you need to think about what it is you do that is creative, thought based, and relevant to the company. It's all too easy to box yourself in by defining your role as something repetitive, overspecialized, and easily replaced (see the earlier discussion).

A brief aside here about architects. Over the past 10 years it has become popular to give senior individual contributors the title of architect. I am sorry, but architects design buildings, not software, not systems, and not networks. I actually worked with a group in which receiving this title was a source of great humor, rather than pride, and that's the kind of group that KV likes to work with. Usually I find it's easy enough to co-opt the language of the management ladder. You're a junior X or a senior X or a director of X or a VP of X. If you want to point out that you're not managing any people, then put in technical, as in senior technical network specialist. Specialist is another good generic word that says you have a defined role but one that's not too tightly defined.

The final area I want to talk about is proving your worth to the organization that you work in. Any field and systems administration that falls into this category that is responsible for the smooth, day-to-day running of an operation suffers from two significant handicaps right from the start.

The first handicap is that people expect things to "just work" without understanding what it takes to keep a set of systems running such that they appear to be always available. The only time people notice you or your group is when something breaks. Then suddenly they're all up in arms and screaming about how they can't get to the web (where they were probably wasting time instead of working anyway), or their particular application is broken, and so forth. I am quite sure you've experienced this problem already, even as a budding systems administrator. Of course, randomly unplugging network cables, waiting for the phone to ring, and then plugging them back in might be an amusing way to make sure that people understand your worth, but even I can't really recommend this course of action.

The second handicap suffered in the systems administration field is that most people in the business do not correctly perceive the worth of your work. The programmers and engineers often get kudos for making their code work and getting the project, whatever it is, out the door, but the role that is played by systems administrators in making sure that all those programmers are productive is rarely recognized, even by programmers themselves, who often think, "Who the hell are they?" and look down upon "supporting" groups such as the sysadmins. This kind of dynamic is akin to drivers of expensive cars complaining about the people who build and fix the roads. It takes a road to drive a car, and you should be thankful for good roads. People who use your systems ought to be thankful when they receive good service, but usually they aren't.

Both of these handicaps need to be addressed in roughly the same way: through communication. While it's vitally important to communicate problems and outages, these should not be the only things that users learn about from the systems administration group. Whenever a new system comes on line or a new service is successfully rolled out, that fact should also be noted, and not in that horrifically saccharine-sweet way so often favored by the HR department. You're not celebrating little Annie's birthday, after all; you're informing your users that their work just got easier. A simple one-page e-mail, stating clearly what was changed and why it's better, is all that's necessary.

If you can remain interested and knowledgeable in a broad set of topics, help to define your own role, and communicate to your users just what it is you do and why it's important to their day-to-day lives, you will definitely lower your risk of becoming obsolete. And all this advice goes for just about everyone in a technical field. Now...what was my password again?

KV

5.10 With Great Power...

> *The possession of great power necessarily implies great responsibility.*
>
> William Lamb, British M.P. 1817

A naive person believes that they require complete control over a system at all times to get their work done, but when it comes to security, this is actually the opposite of what they might want. The best security people I've worked with know that if you take control over a system, you are also taking responsibility for it, and responsibility is something to be taken very seriously. KV has made a career of dodging such responsibility, because if we learn anything from the politics, it is that what we really want is plausible deniability. KV routinely turns down offers to hold the keys to anyone's systems, for reasons that are laid out in the next letter and response.

Dear KV,

I work in a pretty open environment, and by open I mean that many people have the ability to become the root user on our servers so that they can fix things as they break. When the company started, there were only a few of us to do all the work, and people with different responsibilities had to jump in to help if a server died or a process got away from us. That was several years ago, but there are still many people who have rootly powers, some because of legacy and some because they are deemed too important to restrict. The problem is that one of these legacy users insists on doing almost everything as root and, in fact, uses the `sudo` command only to execute `sudo su -`. Every time I need to debug a system this person has worked on, I wind up on a two- to four-hour log-spelunking tour because he also does not take notes on what he has done, and when he's finished, he simply reports, "It's fixed." I think you will agree this is maddening behavior.

Routed by Root

Dear Routed,

I would like to tell you that you can do one thing and then say, "It's fixed," but I can't tell you that. I could also tell you to take off the tip of his pinky the next time he does this, but I bet HR frowns on Japanese gangster rituals at work.

What you have is more of a cultural problem than a technical problem, because as you suggest in your letter, there is a technical solution to the problem of allowing users to have auditable, root access to systems. Very few people or organizations wish to run their systems with military-style security levels; and, for the most part, they would be right, as those types of systems involve a lot of overkill and, as we've seen of late, still fail to work.

In most environments it is sufficient to allow the great majority of staff to have access only to their own files and data, and then give a minority of staff those whose roles truly require it the ability to access broader powers. Those in this trusted minority must not be given blanket authority over systems but, again, should be given limited access, which, when using a system such as sudo, is pretty simple. The rights to run any one program can be whitelisted by user or group, and having a short whitelist is the best way to protect systems.

Now we come to your point about logging, which is really about auditability. Many people dislike logging because they think it's like being watched, and, it turns out, most people don't like to be watched. Anyone in a position of trust ought to understand that trust needs to be verified to be maintained and that logging is one way of maintaining trust among a group of people. Logging is also a way of answering the age-old question, "Who did what to whom and when?" In fact, this is all alluded to in the message that `sudo` spits out when you first attempt to use it:

5.10 With Great Power...

```
We trust you have received the usual lecture from the local
System Administrator. It usually boils down to these three
things:
    - Respect the privacy of others.
    - Think before you type.
    - With great power comes great responsibility.
```

Most people remember the third item as a quote from the comic book Spider-Man, although it's much older than that, but Stan Lee's version is germane in this context. Root users in a Unix system can do just about anything they want, either maliciously or because they aren't thinking sufficiently when they type. Frequently, the only way of bringing a system back into working order is to figure out what was done to it by a user with rootly powers, and the only way of having a chance of doing that is if the actions were logged somewhere.

If I were this person's manager, I would either remove his `sudo` rights completely until he learned how to play well with others or fire him. No one is so useful to a company that they should be allowed to play God without oversight.

KV

5.11 The Letter

> *We don't care. We don't have to.*
> *We're the phone company.*
>
> Ernestine, played by Lily Tomlin

The *we've lost your data* letter is now so common as to be junk mail to most of us, unless we think we can join a class action against the company. For those working in computer security, though, these letters run the gamut from humor, although of the schadenfreude variety, to useful teachable moments. By now it's probably possible to create an entire one-semester course in computer security where each week, instead of reading a quality research paper, the class has to read one of these apology letters and then work backward to figure out what the root cause of the security issue was.

The following letter and response cover a specific, underlying security issue but the continuing increase in the rate of data breaches, and their associated, disingenuous, apology letters belies a lack of actual care about data on the parts of those companies that collect and use it every day. It is just this cavalier attitude toward people's personal information that has led to the creation of GDPR, one of the most sweeping pieces of data privacy legislation ever enacted. GDPR was enacted in Europe, and some might cynically say that it is aimed at reducing the hegemony of the companies that act as personal data vacuums in the United States, such as Facebook, Apple, Amazon, Netflix, and Google, which turn out to form the very appropriate acronym FAANG. There is surely some level of anti-U.S. cynicism in GDPR, but the idea, on the whole, is not wrong. The only entities large enough to hold the FAANG or any other large companies accountable for their data privacy actions are governments, and the power of the courts. Let's be clear, in this arena there are no paragons of virtue on any side; in fact, there are very few who could be seen to have any sort of virtue, but the fact remains that allowing companies to police themselves has been a disastrous failure. When companies are allowed to self-police, they simply hide the breaches until some insider leaks the issue to the press. The reaction to GDPR has been for companies to contract out ridiculous sums and numbers of hours to make sure they can cover their asses when they're found out because nothing combats a cynical attack on a company like a company's cynical attempts at CYA. The true test of data privacy laws will be if they make any dent in the number of apology letters we receive in future years.

5.11 The Letter

Dear Readers,

I recently received a letter in which a company notified me that they had exposed some of my personal information. While it is now quite common for personal data to be stolen, this letter amazed me because of how well it pointed out two major flaws in the systems of the company that lost the data. I am going to insert three illuminating paragraphs here and then discuss what they actually can teach us:

"The self-described hackers wrote software code to randomly generate numbers that mimicked serial numbers of the AT&T SIM card for iPad called the integrated circuit card identification (ICC-ID) and repeatedly queried an AT&T web address."

This paragraph literally stunned me, and then I burst out laughing. Let's face it, we all know that it's better to laugh than to cry. Unless these "self-described hackers" were using a botnet to attack the web page, they were probably coming from one or a small number of IP addresses. Who, in this day and age, does not rate limit requests to their web sites based on source IP addresses? Well, clearly we know one company that doesn't. It's very simple: if you expose an API and a URL is an API when you're dealing with the web, then someone is going to call that API, and that someone can be anywhere in the world.

A large company doing this is basically begging to be abused: it's not like you're just leaving your door unlocked, it's like a bank letting you try 1 million times to guess your PIN at the ATM. Given enough time, and computers have a lot of time on their hands, you're going to guess correctly eventually. That's why ATMs *don't let you guess* a million PINs! All right, in this case the company was not going to lose money directly, but it certainly lost a good deal of credibility with its customers and, more importantly, possible future customers. Sometimes brand damage can be far worse than direct financial damage.

Now we come to the next paragraph, in which the company admits to not having proper controls over its own systems:

"Within hours, AT&T disabled the mechanism that automatically populated the e-mail address. Now, the authentication page login screen requires the user to enter both their e-mail address and their password."

"Within hours?!" Are you serious? At this point I was laughing so hard it hurt, and my other half was wondering what was wrong with me, since I rarely laugh when reading the mail. The lesson of this paragraph is to always have the ability to kill any service that you run and to be able to either roll forward or roll back quickly. In fact, this is the argument made by many Web 2.0, and 1.0, and even 0.1 proponents: that, unlike packaged software, which has release cycles measured in weeks and months, the web allows a company to roll out changes in an instant. In geological time, hours might be an instant, but when someone is abusing your systems, hours are a long time—long enough, it seems, to acquire several hundred thousand e-mail addresses.

Finally, in the next paragraph we find that someone at AT&T actually understands the risk to its customers:

"While the attack was limited to e-mail address and ICC-ID data, we encourage you to be alert to scams that could attempt to use this information to obtain other data or send you unwanted mail. You can learn more about phishing at `www.att.com/safety`."

I somehow picture a beleaguered security wonk having to explain, using very small words, to overpaid directors and vice presidents just what risk the company has exposed its users to. Most people now think, "E-mail address, big deal, dime a dozen," but of course phishing people based on something you know about them, like their new toy's hardware ID, is one of the most common form of scams.

So, some simple lessons: rate limit your web APIs, have kill switches in place to prevent abuse, have the ability to roll out changes quickly, and remember to hire honest people who can think like the bad guys, because they are the ones who understand the risks.

One other thing is for sure, this letter's a keeper.

KV

5.12 The Tickets That...

> *In the beginning was the word and the word was [bull$#!+].*
>
> William S. Burroughs, *The Ticket That Exploded*

Systems that support software development are often some of the worst systems to use, which I am sure says something deep and troubling about those of us who write software. Ticketing systems come in for some of the worst abuse, with most developers asserting, "They all suck." The fact is that most of us have ticketing systems at our jobs, but making people use them effectively, that requires a little extra mojo.

Dear KV,

Have you ever noticed that when you submit a ticket to a ticketing system, no one ever reads it? They just e-mail, call, or drop by your desk to ask you about the ticket, but it quickly becomes painfully obvious that they haven't read what you've written, except for, maybe, the summary? How do you deal with such people?

Ticked off at Tickets

Dear Ticked,

How do I deal with those people? Well, I have this special chair near my desk, and when they sit in it, it runs about 1,000 volts AC through their bodies. I got the idea from an old James Bond film. The problem is, I can't get our janitorial staff to take out the bodies, so I have to do it myself, and the stench is amazing!

OK, no, I don't really electrocute people at work. I do find myself amazed, however, when people who are clearly capable of reading simply don't take the time. The reason you have a ticketing system is to streamline work, and their abject stupidity and neediness pretty much ruins that streamlining. They also seem to interrupt at exactly the wrong moment, sort of like those waiters who always ask you how your meal is when your mouth is full.

I admit that I take a fairly sarcastic, or perhaps a more sarcastic, tone when dealing with such people. I was recently called by a tech-support line that was supposedly fixing a computer for me, and when it became obvious that the caller had not read any of the things that the previous technician had written down, I began to speak very slowly and very quietly and told...them...in...very...short...sentences...what...had...passed...before. I then finished up the call by asking, in what can only be called an icy voice, "Now did you write that down?"

My favorite example of this phenomenon involved a tech-support person I worked with who was dealing with a particularly difficult customer. The product we worked on was an operating system that systems integrators would extend in building their products. Customer accounts were assigned to individual customer support reps, and this rep had a doozy of an account. The customer in question would never read the manual but would call this tech-support engineer, sometimes more than once a day, and ask simple questions that were clearly answered in the manual.

One day the tech-support engineer had his door open and we all heard him say, "Do you have volume X of the manual? Yes? Good. Please turn to page Y. Now read along with me..." and he made the customer read the manual page along with him. I'm not sure if this cured the customer, but it made the rest of us laugh pretty heartily.

I think the best thing you can do when people ignore your tickets is to ask, very politely, "Did you read the ticket?" If it continues to happen, then you can go down the politeness scale to, "Did you read what I wrote?" and from there to, "Can you read?" I would save that last one, though, because from there, there's no going back.

KV

5.13 Of Screwdrivers and Hammers

> *There's nothing more permanent than a temporary hack.*
>
> Kyle Simpson

A common problem is an over reliance on a particular tool, even after that tool is no longer effective, or is even damaging, in the work we're doing. The attachment to a particular tool or method can blind us, and in the case of using a screwdriver like a hammer, quite literally. Some kidding aside, it's important for us to use the best tool for each job, whether that tool is a language, bug tracker, version control system, or even development methodology. Good carpenters don't use screwdrivers to drive nails, just as good koders don't use COBOL for embedded systems programming.

Dear KV,

My employer recently deployed a system on its network that is very sensitive to variations in network traffic. Although our team let people know that the amount of load on our network might cause problems with this particular application, it was decided to deploy the software anyway and see what happened in production. As you can imagine, most of the time things work pretty well; but occasionally, often because of random misconfigurations or because another application abuses the network resources, our shiny software fails completely, resulting in angry e-mail threads and finger pointing. At this point, there is no way to turn back, and we now live in fear of the next time someone adds a new application in the network. There are ways to work around these issues, but people seem unwilling to do the necessary work and are only interested in our group "just fixing the code." Of course, we can patch and hack the code to work around temporary problems in the network, but that doesn't really address the problem. Why is it so difficult for people to understand when they are using a tool the wrong way?

Wrong Way Round

Dear Wrong Way,

Whenever I see people taking one tool and using it, usually poorly, for the wrong job, I am always reminded of screwdrivers. You can use a screwdriver to drive screws, yes, but you can also turn the screwdriver around and use the handle as a hammer to drive nails. Of course, doing this means that you're at risk of poking your eye out, but, you say, "I only need to drive this one nail, I'm sure it will be OK." And it is OK, until the day when it isn't. Software, being far more malleable than a screwdriver, is subject to this extension problem far more often than physical tools.

There are a couple of ways to make your point in these situations. One is simply to let the code break and watch people suffer. I recommend against developing an evil laugh or learning to cackle, as that will give you away. While this is an enjoyable fantasy, it's not very practical in a work environment. There is probably a good reason for your company to use the code you're complaining about, and it behooves you to do what you can to help them use it correctly.

Instead of screaming or cackling or pulling your hair out, you can try to explain to one person, rather than to a group, how the software works and its limitations. If you can find one other person who understands the problem, that can help you in two ways. First, it will make you feel less crazy; there is nothing worse than being the only person who sees or understands a problem. Second, it will help convince others of the correctness of your position. If you can get momentum behind your idea, then maybe you can convince the powers that be to use the system correctly and within its design parameters. Failing that, at least you'll have someone to commiserate with over a beer when the system collapses again.

Like so many problems in computing, the screwdriver problem is a human problem and not a technical one, and thus it requires a human solution.

KV

5.14 Security Reviews

> *This system is too complicated to have an architecture.*
>
> Engineer at a major corporation
> during a security review
> circa 2004

The term *security* in technology has become one of those words that all of us love to hate, for a variety of reasons. On the one hand, it is now well-understood that nearly all software systems are insecure at some level and certainly don't measure up to the world of physical security. On the other hand, it is often used as a catchall to justify often questionable decisions: "Why can't my contractor access the source code?" "Security!" "Why do I have to go through three hoops to read our internal documentation?" "Security!" If you preface "security" with "cyber," it seems that you now have an excuse to print money as well as wear chic, black, clothes ala the Matrix movies.

Looking past the incredible hype that surrounds computer security we come to the more practical and pragmatic questions about how one might evaluate the security of a system, which is really the crux of the matter. While terms like "nation-state actor," "advanced persistent threat," and "Dark Web," might send chills up and down the spines of those who have little real-world experience of computers, or security, the real work remains the drudgery that most people need to but often fail to do to secure real-world systems. The three real pieces of work that one can do to protect systems come down to updates, tooling, and reviews. As the next letter and response discuss the review component, let me lay a few words on you about the first two, updates and tooling.

By far the most commonly exploited computer security flaws in the world today are based on bugs that have already been found, and often fixed, in existing and deployed systems. The problem isn't just the flaw itself, but the fact that even though there is a known fix, a large number of installed systems have not had the fix applied. The physical world equivalent would be for a car company to admit it had an issue with the brakes in your car and issued a recall, but you're just too busy to have the brakes fixed. Because computer systems and their failures are most often hidden from plain view, when we have a computer security issue, you don't, yet, see hunks of shattered metal at an intersection. This inability to see the issue means it's easy to ignore. Sometimes people find another reason not to update their systems, e.g., the fact that an update to one part might break another. As a rant on the fragility of computer systems is an overall theme of this book, I'll not go into that here, but it is the most often given excuse to refuse a software update, that some other program on the system will break if we update the operating system or some set of libraries. The fact is that the best protection against a large number of computer security issues is to track and apply software updates as soon as they are available. The problem of updates has now become such a persistent trope in computer security that many systems are now being designed to not give the user a choice about software updates; they are simply installed whether the user wants them or not, and often they will not even know that the update has been applied. As a koder who believes that the user's will should not be violated, since I'm a user as well as a developer, the idea of forced updates, even for security issues, is particularly galling. Alas, history has shown time and again that allowing people to decide for themselves when it's time to update their software is a losing game.

From the koder's standpoint our best line of defense against security issues is to always write perfect software, without bugs, and to design in perfect security. Once you have stopped laughing... it's not that we shouldn't design and implement our systems for security, but it's pretty clear that despite the best efforts of thousands of people over several decades, the idea that we'll have secure systems just through the use of better practices is laughable, and we need to deploy other techniques to address the problem. Tooling should be our next line of defense against security issues in software after good architecture and coding practices. Static as well as runtime analysis tools are a useful next line of defense against bugs, and since most computer security problems arise from implementation bugs, it is logical to use bug hunting tools in order to find and eliminate security problems in software. Like any piece of software, security analysis tools are of varying quality and require time to learn and apply correctly. A common failing is to purchase such a system, run it against a piece of software, and wind up with a ton of false positive results that then convince the koders involved that the tool is the problem and that their software doesn't need such tools run against it. Anyone who has any inkling of how a static analyzer works knows that it will throw off false positive and false negative results and that it will have to be tuned to the software against which it runs. The true advantage of these systems comes after the tuning, because that's the point at which it can be run nightly or weekly against a large, ever-changing code base and where it can more easily pinpoint bugs that creep in and that might be the source of future security issues. If only more people would take the time to do this right, they'd be able to sleep a bit more comfortably at night, or, in as in the case of KV, in the afternoon.

While updates and tooling are two prongs of a good security strategy, it is the first prong, discussed in the letter and following response, that I have seen be the most productive over the years. Security reviews, like design and code reviews, make it possible to bring the human element, and human intelligence, to bear on the problem of computer security. Good security reviews always bring out the issues that are hidden in the minds of the designer and developer, much in the way that the *stupid human trick* (see Section 5.21) allows a koder to figure out a bug by talking about it out loud. Having to explain your system and its security to other people forces you to surface your assumptions, not just to the review team but to yourself as well, and this type of self-discovery can often help you to find the places where mistakes are being made and security is weak.

5.14 Security Reviews

Dear KV,

I'm working on a project that has been selected for an external security review by a consulting company. They are asking for a lot of information but not really explaining the process to me. I can't tell what kind of review this is—pen (penetration) test or some other thing. I don't want to second-guess their work, but it seems to me they're asking for all the wrong things. Should I point them in the right direction or just keep my head down, grin, and bear it?

Reviewed

Dear Reviewed,

I have to say that I'm not a fan of keeping one's head down, grinning, or bearing much of anything on someone else's behalf, but you probably knew that before you sent this note. Many practitioners in the security space are neither as organized nor as original in their thinking as KV would like. In fact, this isn't just in the security space, but let me limit my comments, for once, to a single topic.

Overall, there are two broad types of security review: white box and black box. A white-box review is one in which the attackers have nearly full access to information such as code, design documents, and other information that will make it easier for them to design and carry out a successful attack. A black-box review, or test, is one in which the attackers can see the system only in the same way that a normal user or consumer would.

Imagine you are attacking a consumer device such as a phone. In a white-box situation, you have the device, the code, the design docs, and everything else the development team came up with while building the phone; in a black-box case, you have only the phone itself. The pen-test idea currently has credence in security circles (which I ascribe to the tittering 12-year-olds who get off on saying they're responsible for "penetration testing"), but, candidly, that is just a black-box test of a system. In point of fact, the goal of any security test or review is to figure out if an attacker can carry out a successful attack against the system.

Determining what is or is not a successful attack requires the security tester to think like the attacker, a trick that KV finds easy, because at heart (what heart?) I am a terrible person whose first thought is, "How can I break this $#!+?" Security testing is often quite easy because of the incredibly low overall quality of software and the increasingly large number of software modules used in any product. To paraphrase Weinberg's Second Law, "If architects designed buildings the way programmers built programs, the first woodpecker that came along would destroy all of society." The difficult parts of security work are constraining the attacks to those that matter and getting past those koders with a modicum of clue who are able to build systems that at least resist the most common script kiddie attacks.

Your letter seems to imply that your external reviewers are interested in a white-box review since they are asking for a great deal of information, rather than just taking your system at face value and trying to violate it. What to expect from a white-box security review, at least at a high level, should not be a surprise to anyone who has ever participated in a design review, as the two processes should be reasonably similar. The review would work in a top-down fashion, where the reviewer asks for an overall description of the system, hopefully enshrined in a design document (please for the love of God have a design document); or the same information can be extracted, painfully, through a series of meetings. Extracting a design in a review meeting takes a great deal longer in the absence of a design document but, again, looks similar to a design review. First, there must be a lot of coffee in the room. How much coffee? At least one pot per person, or two if you have KV in the room. With the coffee in place, you need a large whiteboard, at least two meters (six feet) long. I also suggest implements of torture, or at least a riding crop, to keep people in line.

Then we have the typical line of interrogation: "What are the high-level features?"; "How many distinct programs make up the system?"; "What are they called?"; "How do they communicate?"; and for each program, "What are the major modules of this program?" KV once asked a software designer after he had filled a four-meter whiteboard with named boxes, "What's the architecture that holds all this together?" to which the answer was, "This system is too complex to have an architecture." The next sound was KV's glasses clattering on the table and a very heavy sigh. Needless to say, that piece of software was riddled with bugs, and many were security related. It is not every day that KV wants to switch from coffee to gin and tonic at noon, but then there are those days.

A good reviewer will have a minimal checklist of questions to ask about each program or subsystem, but nothing too prescriptive. A security review is an exploration, a form of spelunking, in which you dig into the dirty, unloved corners of a piece of software and push on the soft parts to see if they scream, or spit green ichor, which burns—it burns and you can't wash the damned stuff off! Overly prescriptive checklists always miss the important questions. Instead, the questions should start broad and then get more focused as issues of interest appear—and trust me, they always will.

When issues are found, they should be recorded, though perhaps not in an easily portable form, since you never know who else is reading your ticketing system. You want to get inside a system and go read the bugs. If you have a bad apple or two inside the company (and what company is free of rotten apples?) and they do a search on "Security P1," they're going to walk away with a lot of fodder for zero-day attacks against your system.

Once the system and its modules have been described, the next step is to look at the module APIs (application programming interfaces). You can learn a lot about a system and its security from looking at its APIs, though some of what you will learn will never be able to be unseen. It can be pretty scarring, but it has to be done. I feel most of these steps ought to have wine (or something stronger) pairings. For readers in California, I recommend a nice indica for this kind of work.

The APIs have to be looked at, of course, because they show what data is being passed around and how that data is being handled. There are security scanning tools for this type of work, which can be used to direct you toward where to perform code reviews, but it's often best to spot-check the APIs yourself if you have any type of ability or intuition around security.

Lastly, we come to the code reviews. Any reviewer who wants to start here should be fired out of a cannon immediately. The code is actually the last thing to be reviewed—for many reasons, not the least of which is that unless the security-review team is even larger than the development team, they will never have the time to finish reviewing the code to sufficient depth.

Code reviews must be targeted and must look deeply at the things that really matter. It is all of the previous steps that have told the reviewers what really matters, and, therefore, they should be asking to look at maybe 10 percent (and hopefully less) of the code in the system. The only broad view of the code should be carried out, automatically, by the code-scanning tools previously mentioned, which include static analysis. The static analysis tools should be able to identify hot spots that the other, human reviews have missed, and then the humans have to go back into the dark corners of the code and again try to avoid being sprayed with green ichor.

With the review complete, you should expect a few outputs, including summary and detailed reports, bug-tracking tickets that describe issues and mitigations (all while being secured from prying eyes), and hopefully a set of tests the QA team can use to verify that the identified security issues are fixed and do not recur in later versions of the code. It's a long process littered with broken hearts and coffee mugs, but it can be done if the reviewers are organized and original in their thinking.

KV

5.15 Getting Back to Work

> *I'd rather dig ditches; everyone needs ditches.*
>
> Anon

How does one scratch the itch to get back into coding when one has gotten out of the habit? As KV points out in the following letter and response, your best bet might not be to bother doing that at all. The allures of koding are well-known: fame, fortune, beautiful people throwing themselves bodily at you when they figure out that *you* know how to make computers do what you will them to do. Koding, like most things, takes study and practice, and the real question being asked here isn't how to get a job, but how to get back into the practice of programming.

Dear KV,

I am an IT consultant/contractor. I mainly work on networks (have my CCNA) and Microsoft OSs (MCSE). I have been doing this work for more than eight years. Unfortunately, it is starting to bore me, and there is less and less job satisfaction in my life.

My question is: how would I go about getting back into programming? I say "getting back into" because I have some experience with programming. In high school I took two classes of programming in Applesoft BASIC (archaic, I know). I loved it, aced everything, and was the best programming student the teacher ever saw. This boosted my interest in computer science, which I pursued in college.

In college, I took classes in C++, Java, and web development (HTML, XHTML, JavaScript). I did great and had fun also. For various reasons, I ended up leaving college and becoming a network administrator, and for the past eight years have been doing this and that and everything else IT related. But I haven't been programming. The extent of my programming work experience is with MS Excel macros (yuck!) and basic VB coding in MS Access.

So how could I start becoming a programmer? Visual Studio 2005? Java? Eclipse? I enjoy self-learning and have found that achieving certifications gets my "foot in the door." Is there a particular suite that I could certify in to get my newly desired career going?

Thanks in advance.

Jonesing for a Job

Dear Jonesing,

Wait, you've gotten so good at your job that you're bored? Why not take up golf or painting? Perhaps download more "content" from the Internet! Why on Earth would you want to give up a job that you know just to become a programmer?! What could possibly be possessing you? Don't you know that programmers put in long hours to meet impossible deadlines all to make a bunch of rich guys richer?

OK, if you've gotten this far, then I guess maybe I should actually answer your question. There are as many answers to your question, of course, as there are programmers. How to go about moving into a coding job depends on a few things. The first is, "What do you like to do?" There is no point in working hard to learn a language or a system if you don't enjoy it; you'll just wind up right back where you are now, wanting to do something else. Find the kinds of problems you like to solve, then see how they're solved today, and see if that's the kind of thing you like to do. Since you say you already have a computer science background and already know a couple of programming languages, I don't really see much reason to go for any special certifications; I have never seen the point of most certifications as they only certify that you can pass a test, not that you can think about problems, which is the more important skill. I also don't think you should worry about which set of tools you use just yet; there are other things to work through, like step 2.

The second step, and be thankful that there aren't 12 of them, is to pick a project. For me, I can't learn anything unless I have a project around what it is I want to learn. Try to pick something you think you can actually do. "Write an operating system," though it's a fun goal, and, actually possible to do, is likely not the right place to start. Working on an open source project, like FreeBSD, Apache, or Open Office, might be another way to get started. Find something you need to use or work with on a relatively frequent basis and try to extend or fix it. Most open source projects have long lists of open bugs; pick a few of those and try to fix them and submit patches to the maintainers.

Lastly, take every opportunity you can to learn about your new area. That doesn't mean spend lots of money, or con your employer to spend lots of money, on classes and conferences. Of course, if you find a class in Hawaii and they're willing to send you, well, I can't argue with that, but I can't call it learning either. Find the journals, magazines, and web sites that cover your newfound area of expertise and read them regularly.

Once you think you have what it takes to start your new career, look at some entry-level positions that will allow you to learn more and apply for them. Be prepared to take a pay cut because moving from eight years of experience in network administration to entry-level work programming isn't likely to be a big monetary win first off, but then you're doing this for the fun and challenge, right?

KV

5.16 Open Source Licenses

> *The GNU GPL was not designed to be open source.*
>
> Richard M. Stallman

Computer science, like most sciences, has always had, along with its commercial aspirations, a high level of volunteerism. Long before the term "open source" was codified, and of course now incessantly argued over, programmers wrote and shared code with each other in many ways and mediums. The early micro-computers were made that much more interesting because of computer magazines and computer clubs where participants could share the code they wrote and which they put into the public domain, which was effectively giving the code away, without a license or any request of credit or attribution. The mainframe and mini-computer eras had their own exchanges of software, including SHARE, a user group founded in 1955 that continues to this day. Often times it was easy to share software in the early days of computing because the companies involved saw no value in software, as they made their money on hardware and repairs, and sometimes because, as people do, it was just fun and intellectually interesting to show people the cool thing you'd done, and you'd be happy if they wanted to use your code to do something else, bigger and cooler than you had done on your own.

I recently acquired, through a friend, my favorite computer from that era, an Amiga, which is what I hacked on in college and the system on which I wrote my first piece of commercially sold software. Four large boxes arrived; only one had the computer, and the other three were books and software, a small amount purchased, but most of the 3.5" floppy disks (yes, kids, I'll wait while you look that up) contained software with no licenses at all, just hundreds of small and large programs given away free by their authors, to amuse, entertain, and make productive other people who shared their interests.

With the advent of open source licenses we attempted to codify what we meant by sharing, which might seem a bit strange since the definition of sharing something seems relatively clear, but the products of the mind are a bit harder to codify than, say, an apple. Sharing an apple, a finite resource, is easy to see, which makes it harder to argue about. Sharing an intellectual resource that can be copied and applied repeatedly, infinitely, carries challenges that have now kept real lawyers, as well as koders who think they're lawyers, busy for nearly four decades, and given the number of new licenses that continue to pop up, there is no sign of stopping.

Dear KV,

I run a small set of web servers at home, hosting web pages for friends. I've been building a small package for analyzing web logs over the last few months, in my "spare" time of course, and now I want to post it on SourceForge. SourceForge requires you to place an open source license on the code, and I was going to put it under the GPL (Gnu Public License), but then I read about all the fighting over the GPL3, and I was wondering if this would affect my package. I did a search for other licenses, and it seems that everyone has their own. How do you decide which one to use?

Unlicensed

Dear Unlicensed,

First, let me state, right up front, I am *not* a lawyer and have never played one on TV. I know some lawyers, purely on social, well, really drinking, terms, but that's the extent of my legal knowledge. With that caveat out of the way, which will hopefully placate the lawyers at Queue, with whom I have never had a drink, let me try to answer your question.

Personally, I use the BSD license, also known on the opensource.org site as the ¡href a="http://www.opensource.org/licenses/bsd-license.php" ¿NewBSD¡/a¿ license. I don't just use the BSD license because I'm an old cranky BSD coder, though I am, but because it gives me and my code the greatest amount of freedom with what I perceive to be the least amount of risk. I believe most programmers want people to use their code and to feed back patches and to stay out of legal trouble. All well and good, and that's what the BSD license gives you. The GPL, well, that's another kettle of hot boiling oil.

The best way I can describe the GPL is that it's the roach motel of licenses and code checks-in, but it doesn't check out. Putting the GPL on your code means that not only are you sharing it, but anyone who uses your code must share with you. Enforced sharing has always seemed to me like saying, "I'm taking my marbles and going home." You see, the part of the GPL that's problematic to me is that if I use your code, then I have to give you my code if I ship a product with your code in it. What if I am willing to give you only part of my code but for some reason I don't want to give you all of it? You can't do that with the GPL. In reality, the GPL is the Scylla to Big Corporate's Charybdis, and for those of you who forgot Greek mythology it puts you between a rock and a hard place. Big Corporate licenses are completely closed, you use the code on their terms, but the GPL is similar in that they're dictating what you do with your code, the code you wrote, because it's an extension of theirs. Most people call this a "viral" license, and many projects that don't use the GPL are very careful to wall off GPL code from the rest of their code to prevent the virus from poisoning them.

There is also the route of putting your code in the public domain, which provides you no control and theoretically no protection. I've not done that since I started using the BSD license because with the BSD license you're also saying, "Don't sue me! This has no warranty!" which are good things to say. You don't want someone integrating your code into their product and then looking to you for damages when they go out of business.

It's my belief that many people use GPL because it's famous, not because they actually want to force anyone who touches their code to share with them. On numerous occasions I've contacted authors of libraries I wanted to extend to ask them if they could not use the GPL, and usually they said, "Sure, why?" Once they understood why, well, they usually changed to something less viral, sometimes even BSD. Intellectual laziness of this sort is the very thing that leads people to use technologies for projects for which they are totally unsuited, like putting a mission-critical system on Windows, but I digress...

Of course, don't take my word for it; ask a lawyer if you're truly paranoid.

Just so you know, "THIS ARTICLE IS PROVIDED BY THE COPYRIGHT HOLDERS AND CONTRIBUTORS "AS IS" AND ANY EXPRESS OR IMPLIED WARRANTIES, INCLUDING, BUT NOT LIMITED TO, THE IMPLIED WARRANTIES OF MERCHANTABILITY AND FITNESS FOR A PARTICULAR PURPOSE ARE DISCLAIMED."

KV

5.17 So Many Standards

> *Turn him to any cause of policy,*
> *The Gordian Knot of it he will unloose,*
> *Familiar as his garter*
>
> Shakespeare, *Henry V*

It is very easy to be cynical about standards, not because there are so many but because they are often used as cudgels with which we think we can get our way when purely technical arguments fail us. While there are definitely cases of competing standards, which are put forth by companies to control some segment of a market and to exclude others, there are also well-written standards that, when adhered to, bring about a far better system than one that was cobbled together with less thought. While in Chapter 4 we discussed network protocol standards, including how they're used and misused, here we're talking about a very personal standard, the coding standard. The rise of distributed software development, through large open source projects, has made this discussion all the more important, because if you think it's hard to get five koders at one company to produce code to a coherent standard, try doing that with 500 koders, spread across a planet, many of whom do not speak a common tongue. Coming to some sort of agreement on this type of standard, early in a project, is perhaps the most important nonalgorithmic decision a set of koders can make. The fact that there are so many standards to choose from can be a blessing or a curse. Having many choices means there ought to be one that will fit a project, but it also means that people can, if not properly led, argue about them ad nauseum.

Dear Kode Vicious,

We're starting a new project at work and are trying to decide on a coding standard. Of course everyone on the team, there are 10 of us, wants to use their own personal favorite or the one they used in their last company. How do we choose one, and does it really matter?

So Many Standards

Dear SMS,

Let me ask you, just how much time has your team spent deciding on a coding standard? A day, a week, a month? Don't you people have anything better to do?! I mean come on, we've all been over this time and again, and no one likes the answer, but an answer needs to be had. Get together, get some beer, have everyone bring along their favorite coding standard. Then lock the door, and no one leaves until you've agreed. Luckily, the beer will make the decision go quicker, because, well, if you lock the door, someone is going to have to go, you know? Well, perhaps that's not the most productive way but at least it gets you a reasonably quick result.

The point of a coding standard is so that all the poor bastards you work with will be able to understand, hopefully, what your sleep-deprived minds created at 2 a.m. some Sunday morning. You remember, the code you created the day before a release. This understanding will become necessary at a later date. Perhaps when that first bug report comes in, about a day after the release, at 3 a.m., while you're catching up on the sleep you missed, or better yet, drinking down your most recent paycheck. So just remember, the poor bastard you help could be yourself.

So, if I were you, I'd go for readability, readability, and readability.

A good coding standard:

- Makes correlated chunks of code look like they're related, and not like the offspring of close relatives.

- Uses enough white space that those of us who wear glasses (and you kids out there, you'll be wearing them too, soon enough, so stop laughing now!) can tell which code block goes with which `if`, `while`, or `case` statement. It should also not use so much white space that you need a new 23" monitor just to read the code.

- DoesNotMakeItSoThatNamesAreSoLongTheyLookLikeSentences.

    This practice, in particular, makes me do things that I can't mention here, and which cost a lot of money if you have to pay for them. The idea, like so many ideas, has a kernel of goodness, that the reader of the code should be able to read what it does like written English text. But in the end you can't tell:

ReadInputFileFromDiskIOComplete

from

ReadInputFileFromNetIOComplete

as well as you'd like. Picking rational names is more important than reproducing William Burroughs in a device driver.

- Makes it clear when there is a coding error. If the way the code is written hides things, like accidentally open `else` statements (you know, you thought you added a line of code to the `else` block but forgot the braces?), then it's not helping you.

I say you start from there. The coding standard itself should be only four pages long. Most engineers don't want to read a loose-leaf notebook full of process BS to write code. Each statement in the coding standard must be written so that it is unambiguous (yes, it's a big word, look it up). Think of it as a list of do's and don'ts. Just say how routines are named, how code is indented, where braces are located, and what variables should look like and be done with it. Then get back to work and stop drinking my beer!

Oh, and one thing, I've always laughed at the comment, "What if you get hit by a bus?" My co-workers rarely get hit—by buses.

KV

5.18 Books

> *You don't have to burn books to destroy a culture. Just get people to stop reading them.*
>
> — Ray Bradbury in *Fahrenheit 451*

KV loves books. Well, I must, as I seem to have been stupid enough to continue to write them. I don't just love books, what I really love are good books, and the following two pieces give a list of books that I think are worth reading. The number of books written on computer-related topics has continued to grow over the entire history of koding as a discipline, but the number of good books has been, to my eye, relatively constant. Sturgeon's Law states, "90 percent of everything is crap," and Sturgeon was an author, admittedly of science fiction, but, still, the law applies in many places including books about computing. The books I've listed in these two pieces are uncontested classics in the computing world, and in the years since I wrote these pieces I can only come up with a few more that I'd consider in the same class.

What makes a good computing book is a mixture of the things that make a good book generally, and things that must be present in a computing book specifically. The general qualities are narrative flow and clarity of thought and description. Whether a book is written about compilers, photography, or the history of primates, or is a work of complete fiction, it must present the reader with a narrative. The book must take the reader from a point of ignorance, about the topic, or in fiction about the characters, to one of enlightenment. Too many computer book authors simple splat the pages with facts without connecting those facts to a narrative that makes those facts useful and relevant to their readers. For any type of technical book the next requirement is clarity. The book must avoid the use of jargon, overlong sentences, and many other foibles that are called out in standard books about writing.[2] The point of writing isn't to impress the reader with how smart the author is, it is to impress upon the reader the knowledge that the author is attempting to share.

From the general we can now turn to the specific problems that appear in many computer books, which fall into three categories: examples, cross-references, and indexing. Nearly all of the books that make my list have excellent examples to go along with the text. Sometimes authors make their chapters mostly out of examples, as if showing the reader page after page of code or diagrams will help them to understand what the author is trying to say. Flooding the reader with code is nearly the same as dumping an old-school, green bar listing on their desk of the code with few comments and asking them to read it. Examples need to be relevant to the text, and the text must come first and not be added as a sort of fat footnote to the example. Cross-referencing is important

---

2. William Strunk/E. B. White: The Elements of Style, Boston (USA) 1999.

because most people do not read computer books like novels, front to back. Computer books are often used as reference works, and for a reference to work it has to have coherent cross-referencing, not that every sentence must point off in three directions like some bastard form of Wikipedia, but the cross-references are there to make it possible for someone to look for a concept in Chapter 5, and, if they need it, be confidently pointed back to Chapter 2, when they find that the concept in Chapter 5 depends on Chapter 2. Finally, we come to the index, which for a reference work is consulted far more often than the table of contents. Great computer texts have well-built indices, because it is here that the reader will nearly always go, after their first read of the book, to find just that concept that they, right now, can't put their finger on.

Dear KV,

I just finished my degree and started a new job at a big IT company in Silicon Valley. The work is OK, if a bit boring; how much can one do with web pages, and who cares about blogs anyway? I took this job because the company develops all their code on open source systems, and that means I get to look around at the code while fixing menial bugs, which is what they pay me for. Most of the people in my engineering group are recent grads, and a few of them seem interested in more than just taking home a big paycheck, counting their stock options, and planning on which car they'll buy when their first stock chunk becomes available. A few of us are passing around our favorite tech books, and we also occasionally pass around your column. We have a little lunch bet going on about which books you would recommend, if any, for your readers to have on their bookshelf. The person in the group whose list most matches yours gets lunch bought by the rest of us.

So, what's on your list? I want that lunch!

Hungry Reader

Dear Hungry

What do you mean you "occasionally" pass my column around? I expect each and every one of you to go and subscribe to Queue right this instant! If not, I shall have a bunch of rabid copyright lawyers visit you in the middle of the night and do things that would make even my psychiatrist uncomfortable!

OK with that squared away, every programmer and engineer I know, or at least those who are still speaking to me, have a small cluster of books always near to their work area. I think it's that set of books, the ones you can't, or shouldn't, live without that you're after. The problem with book lists is that they're highly subjective and in a space as diverse as IT and computer science, where there is a new fad every, well, there goes one now! Uh, what was I saying? Oh, right; the lists are always subjective and pompous, but then so am I, so, I guess I can give you my list. One thing I like to point out is that not only are these books useful, but they are also well-written and easy to read, which is very important when you have 400 or more pages of complex ideas to read through. There is never any reason to read a book, no matter how important someone says it is, if it is not also a well-crafted piece of writing.

The Art of Computer Programming by Donald Knuth—Perhaps the best-known master works on computer science, these books are both reference and relaxation. I received my first set as a Christmas present my freshman year of college, and yes, I requested them, as Santa rarely peruses the computer science section of bookstores. At first I didn't understand very much in them, and I've never actually read them through, but when you have a question about an algorithm or you're even thinking of optimizing some piece of code, these are the books to spend the day with. Either you will a) find out that Dr. Knuth already knows the answer or that b) no one does and you're on your own. The books have been being written for almost 40 years now and are always worth having near to hand.

The Art of Computer Systems Performance Analysis: Techniques for Experimental Design, Measurement, Simulation, and Modeling by Raj Jain—A book that seems to be much less well known than it should be. First published in 1991, it reads a bit dated now; the hardware used in its examples will either bring a nostalgic tear to your eye or have you just say, "Who is DEC?" Dr. Jain is heavily involved in the networking side of computing, and that shows in this book, but it is much more than a book about networking; it's a great book on applying the scientific method to solving problems in computer science. The book covers such useful topics as proper experiment design, workload selection, and all the other things you need to correctly approach performance problems in your systems.

Anything written by Richard Stevens including, but not limited to, *TCP/IP Illustrated Volume 1 and 2*. Richard Stevens loved to write, and that's very obvious when you read his books. Most of the books he wrote were about networking, and TCP/IP in particular, but some were broader, covering subjects like programming in the Unix environment. Each book is interesting to read, has plenty of relevant examples, and teaches you something on every page.

The Practice of Programming, Brian W. Kernighan and Rob Pike—One of the best books, and a short one to boot. Less than 300 pages and yet filled with interesting stories about programming and filled with practical advice. One of those must-reads, and must-read-agains.

And, finally, a noncomputer book, *The Elements of Style* by Strunk and White. No, not a book on how to dress for those of us who can't figure out if orange and green really do clash but a very short book on the proper use of written English. Why would I suggest such a book? Well, here is my reasoning. If computer science is to be considered a science, then it is important that computer scientists, a group into which I also put all programmers and computer engineers, need to be able to communicate their findings. Science is, after all, the pursuit of knowledge via the scientific method, and one of the important components of the scientific method is that you be able to tell another person what you did and how you did it so that they can verify your work. I don't care how clever your code is; if you can't explain it to someone else, it's almost useless, and so the ability to write in the language shared among most currently working computer people, English, is actually of relative importance. I'm sure my editors wish I referred to this little book more often.

Now, did you get the lunch, and how do I get my cut?

KV

5.19 More on Books

> *Time enough at last!*
>
> Henry Bemis, from the *Twilight Zone* episode of the same name

Before we cover the last book piece, I'd like to update KV's book list with a few more titles that should also be on every koder's e-reader, or shelf, if they're still enticed by paper.

The Practice of Programming by Kernighan and Pike, which is now more than 20 years old, remains terrifyingly relevant. As KV has argued in many of these pieces, it is important to program in the idioms that are relevant to your programming language and environment, and this book goes into some depth on this topic, while remaining readable and enjoyable throughout.

Hackers Delight by Henry S. Warren Jr. is one of those books that you can spend an hour, an evening, or a weekend with. It's a hard book to categorize, but if you're the kind of koder who wants to know not only how the machines work, but some tricks for getting them to do things you want in clever ways, then you definitely want this book, and a pad and pencil, nearby.

UNIX and Linux System Administration Handbook by Evi Nemeth, Garth Snyder, Trent R. Hein, Ben Whaley, and Dan Mackin was foisted on me by a boss early in my career. "What do I need to know about sysadmin? I'm a koder," was probably the politest version of how I asked why I should read this book. This book comes under the rubric of "you will be more efficient in your work if you understand the system at many levels." If all you can do is open an editor and run a compiler, you're far less efficient than if you can diagnose why your network connection is flaky, before calling in your local sysadmin.

Looking through my book list, these are three that I had not mentioned that I recommend to everyone who kodes.

Dear KV,

I read your response to *Hungry Reader*, and I indeed have all of the books you listed except Jain's, which I am going to get. I would include just one more that I feel is indispensable, *The Mythical Man-Month* by Brooks. It's good to get a copy for your manager too. What do you think? Should this one be on your list?

Nostalgic over PDP-11s

Dear Nostalgic,

At the risk of drawing another web comment on being an antique for recommending an older book, yes, your suggestion is a good one. The thrust of my original response was around technical books, whereas *The Mythical Man-Month* is a management book, although one that gives me dry heaves a lot less than most books in the management section of my local bookstore.

I actually remember the book fondly for several reasons. Its conclusions were relatively obvious to anyone who had worked in a software company for more than five minutes, and yet you can use it to beat stupid managers over the head with even today. It has stood the test of time for two reasons; one is that it was well written, an uncommon quality, and because people haven't gotten any smarter about managing large projects in the last 50 years. One other good thing about *The Mythical Man Month* was that it was short enough that I could read it and return it in time to recoup the full price from my university bookstore. I took the money and bought beer for a 4th of July party. No, I am not kidding. So, I don't have my copy anymore, but I do have some fond, if confused, memories.

KV

5.20 Keeping Up to Date

Time isn't the main thing. It's the only thing.

Miles Davis

While books are an excellent way to learn an existing topic, keeping up to date on the latest developments in technology requires one to read more broadly. While many people do this via webinars and watching videos, I find the best way to keep up is to follow a few journals and conferences that have high-quality papers published in them. Since my areas are operating systems, networking, and security, I tend to follow SIGOPS, SIGCOMM, a few USENIX conferences, and the Oakland security conference. As we see in this letter and response, there are easier and harder ways to keep up using academic papers, and for once KV suggests the easy way out.

Dear Kode Viscous,

My boss keeps complaining that I have an NIH attitude and that all my best ideas are already "out there in the published literature," wherever that is. It's really getting me down. Have you thought of any clever ways to get him off my back?

Bummed

Dear Bummed,

Well, I do have my ways of keeping bosses off my back, but I'm not going to share them here. First, because many are illegal, and second, because I think your boss may be right.

It's been my experience that the last time many koders read a paper was when it was a requirement in some course at school. Although there are many books written about the latest buzzwords in the industry, these never cover the basics, the underpinnings of what we do every day, and they also don't talk about major changes in how we build systems. It's all fine and good to learn the language of the week, but if you don't understand how and why languages are written, you'll probably pick the wrong one for your project. The same is true for any part of our field, whether it be databases, web application frameworks, security products, or networking protocols.

"But journals are so boring!" I hear you whine. Well, yes, there isn't a lot of hot bedtime reading out there, except for Queue, of course. I will admit to taking a particularly good set of conference proceedings to bed, but, as you well know, I'm an extreme case.

What it comes down to is that it is your responsibility to be informed about your field. If you were a pipe fitter and you missed the switch from lead to copper, you would certainly be doing your customers and yourself a disservice.

There are a few hints I can give you here to make this process a lot easier. First of all you should ask around and see what your co-workers read. Find someone whose code or designs you respect and see what's on their shelf. Then borrow an issue or two from them, or see if you can read some of the articles online.

Once you have a few different journals, head to a local coffee shop, or bring them on your next plane flight. Commuting is also an excellent time to read, but only if you're *not driving a car!* There are two things you should remember when reading a journal. The first is that most, if not all, journals publish abstracts for all their papers. Read all the abstracts first, and then decide from there which papers you want to read. The second thing to keep in mind is that papers are *not* novels. If you don't like the paper you're reading and you're not getting anything out of it, then put it aside. Life is too short to read poorly written papers.

Now, before I run out of space, I should put in a plug for the good folks here at the ACM. The ACM, of course, has journals and conference proceedings that cover every aspect of the IT industry, and that's a great place to start. My other two favorite sources of journals and proceedings are the IEEE and Usenix. Each organization serves slightly different communities, but they're all worth a look. Now put down that copy of *Snow Crash* and get back to work!

KV

5.21 For My Last Trick

> *Oh Bullwinkle, that trick never works!*
>
> Rocky J. Squirrel

One of the most consistent uses to which you can put your human colleagues and co-workers to is that of the patient listener. KV is not a particular patient listener himself, but he's always happy to find someone who is. The trick mentioned here might not even require a person; you could just as easily put a rubber duck on your desk and explain your problems, but your co-workers might find that a bit too strange. The following response was written to a follow-up letter, but the response stands on its own, I think, and is the final bit of advice I'll leave you with on dealing with, and utilizing, the humans around you.

Dear KV,

I read the comments on Heisenbugs ("Kode Vicious Bugs Out," April 2006) and am surprised you did not mention the approach that tends to work the best in solving these particular bugs. That approach is the random passerby who looks over your shoulder and immediately points out the error.

Passing

Dear Passing,

The debugging practice you're referring to is what I call the "Stupid Programmer Trick." There are actually two different versions. The one you mention is actually the less reliable of the two, because it depends on a chance encounter.

The other version, which I prefer, is where I walk over to a co-worker—it could be anyone, not just an engineer, just someone to stand there and go "uh huh" a lot—and start explaining the problem I'm having. It could even be a marketing person, but then I have to buy them drinks, and that cuts into my own drinks budget.

To use this trick, you start explaining your problem, pointing at the code or diagrams you're using to think about the problem. If you're talking to someone with a clue, you might get lucky, and that person might find the bug, or at least ask you a good question. But at some point, *bang*, you smack your forehead—being bald, I do this a lot; it makes a nice slap sound—and say, "Eureka!" and jump from the bath. No, wait, that was someone else. At any rate, you get that lovely feeling of having found the problem. Thanks for reminding me.

Index

A

abstractions, 129–139
    code reuse, 131
    maintenance of, 130–132
    object-oriented languages versus scripted languages, 133–135
    readability of, 131
    testability of, 130
    utility of, 130
abuse of code, reuse of code versus, 13–15
address lookup, 29–30
AI (artificial intelligence), 245–247
aligning fields, 196–197, 198–199
allocation
    of memory, 4–6
    of ports, 190–192
Apache source base, 66
APIs
    dependency analysis, 140–142
    in security reviews, 282–283
architects, 265
arrangement of code in libraries, 10–12
The Art of Computer Programming (Knuth), 295–296
The Art of Computer Systems Performance Analysis: Techniques for Experimental Design, Measurement, Simulations, and Modeling (Jain), 296
artificial intelligence (AI), 245–247
Asimov, Isaac, 246

assumptions
    documenting, 76, 81
    in testing, 83
attribution when copying code, 39
auditing
    root users, 267–269
    for security holes, 69–70
    sensitive data access, 229
authentication, 149–159
    Digital Rights Management, 152–154
    encryption versus, 149–151
    web browser example, 155–159
automatic documentation, 78–79

B

bandwidth, latency versus, 208–214
bike shed example (nitpicking), 239–241
black lists, 74–75, 164
black-box security reviews, 281
blinding, 178–179
block diagrams, 257–258
book recommendations, 293–298
BPF (Berkeley Packet Filter) packet classification, 29
brainteasers in interview process, 258–259
branched development, when to merge, 96–98
broken builds, 242–244
Brooks, Frederick P., 260, 298
brute-force approaches to code spelunking, 66
BSD license, 288

305

Index

buffer overflows, 198
bug fixes, checking in changes, 18–20
builds, broken, 242–244

C

C++
    C versus, 55–57
    memory leaks, 8–9
    when to use, 47
cache misses, 61–64
career obsolescence, 263–266
CARP (Common Address Redundancy Protocol), 192
centralized version control systems (CVS). *See* version control systems
checking in changes
    comments, 48–50
    size of check-in, 18–20
chess, 246–247
choosing programming languages, 45–47, 55–60
classes, 139
clients, limiting, 217
clocksource.h, 31–33
code profilers, 70
code quality, 58–60
code reuse
    abstractions, 131
    code abuse versus, 13–15
    copying code, 39–41
    IPFW example, 28–30
code reviews, 250, 283
code samples in interview process, 258
code scanners, 117–118
code spelunking, 65–72
coding standards, 290–292
coding style, 1–2
    checking in changes, 18–20
    choosing standards, 290–292
    clocksource.h example, 31–33
    code reuse versus abuse, 13–15
    colorful language, 21–23
    commented-out code, 44
    comments, 43
    constants, 43–44
    indentation, 2
    libraries, 10–12
    naming schemes, 1
    nested files, 16–17
    whitespace, 123–125
coding substance, 2–3
    composeability, 2–3
    conciseness, 2
    copying code, 39–41

correctness, 2
else clauses, 43
forced exceptions, 24–27
garbage collection, 7–9
log output updates, 34–36
memory allocation, 4–6
colorful language in code, 21–23
Colyer, Adrian, 25
commented-out code, 44
comments, 43, 48–50
    code quality and, 59–60
    as documentation, 80
    standards implementation, 232–233
Common Address Redundancy Protocol (CARP), 192
communication failures
    management's lack of understanding, 236–238
    nitpicking, 239–241
compilers, static analysis integration with, 117
composeability
    of code, 2–3
    of systems, 127
computer book recommendations, 293–298
conciseness of code, 2
concurrent systems, 99–101
conditional breakpoints, 68
conditional compiles, 44
conference proceedings, 299–301
configuration data, retaining during upgrading, 102–104
constants, 43–44
control interfaces, 87–88
converting legacy apps, 184
cookies, 158–159, 162
copying code, 39–41
correctness of code, 2
cross-references in computer books, 293–294
cross-site scripting, 160–165
Cscope, 68, 71–72
CSS (cross-site scripting), 160–165
CVS (centralized version control systems). *See* version control systems

D

data privacy, 176–180, 270–272
deadlock, 101
debugging
    code scanners, 117–118
    in code spelunking, 67–69
    conditional breakpoints, 68
    copied code, 39–41
    hardware, 119–122
    in interview process, 258

monitoring system optimization and, 105–108
networked systems, 203–207
rebooting and, 115–116
scaled networked systems, 215–220
scientific method and, 52–54
stepping backward, 68
"Stupid Programmer Trick", 302–303
threaded programming, 147
watchpoints, 68
defined roles, 265
dependency analysis, 140–142
design reviews, 248–251
device drivers, I/O control for, 142
Digital Rights Management, 152–154
digital signatures, 158–159, 179
distributed systems. *See* networked systems
documentation
    automatic, 76–79
    clocksource.h example, 31–33
    formats for, 109–111
    of hardware, 120–121
    what to include, 80–82
Doxygen, 78
DTrace, 72
dynamic analysis of code, 66

E

The Elements of Style (Strunk and White), 296
Eliot, T. S., 253
else clauses, 43
embedded systems, memory management, 7
encryption
    authentication versus, 149–151
    of log output, 179
    in networked systems, 227–229
    null encryption, 91–92
engineering, management's lack of understanding, 236–238
ephemeral ports, 193–195
error handling, forced exceptions, 24–27
examples in computer books, 293–294
extensibility of network protocols, 199
Exuberant Ctags, 71

F

feature updates, checking in changes, 18–20
fields, aligning, 196–197, 198–199
file-sharing systems, 185–187
finding code, 37–38
flags, grouping, 199
forced exceptions, 24–27
Frankenstein (Shelley), 237

functions, calling other functions, 68–69
Futurist Manifesto, 58

G

garbage collection, 7–9
gardening analogy, 11–12
GDPR, 270
Global, 71
global variables, 44
Go, 55
GPL (Gnu Public License), 287–289
gprof, 72
grouping flags, 199
gtags, 71

H

Hackers Delight (Warren), 297
hardware
    debugging, 119–122
    multicore, 99–101
    performance optimization, 61–64
hashes, 151, 158, 178–179, 217
Hein, Trent R., 297
Heisenberg, Werner, 105
"Heisenbugs", 105–108
hiring process, interviewing people, 256–259
Hopper, Grace Murray, 208
hosts, naming schemes, 252–255
hot spots, 217, 218
htags, 71
hypotheses in scientific method, 53

I

IANA (Internet Assigned Numbers Authority), 222, 223
idioms in programming languages, 63
IETF (Internet Engineering Task Force), 221
indentation, 2
indices in computer books, 293–294
infrastructure, broken builds and, 243–244
input validation, 73–75, 160–165
intelligence, 245–247
Internet Assigned Numbers Authority (IANA), 222, 223
Internet Engineering Task Force (IETF), 221
interoperability testing for networking standards, 232
interviewing people, 256–259
I/O control, 142
IPFW (IP Firewall), code reuse example, 28–30
IPv4 packet headers, 191

J

Jain, Raj, 296
Java
    forced exceptions, 24–27
    garbage collection, 7–9
    systems design and, 181–184
journals, 299–301

K

Kernighan, Brian W., 63, 296, 297
kill switches, 271
Knuth, Donald, 295–296
ktrace, 72

L

latency, bandwidth versus, 208–214
layer proliferation, 174–175
learning programming, 284–286
legacy apps, converting, 184
libraries, 10–12, 147
licensing, 287–289
limiting clients, 217
Linux kernel source base, 66
log output
    for hot spots, 218
    nonblocking I/O and, 219–220
    organizing, 116
    rebooting and, 115–116
    for root users, 267–269
    security, 176–180
    updating, 34–36
lossy communication over wireless networks, 225–226

M

Mackin, Dan, 297
magic numbers, 43–44
maintenance, 93–95, 130–132
management
    broken builds and, 243
    understanding of technology, 236–238
marking up standards, 231
memory leaks, 8–9
memory management
    garbage collection, 7–9
    memory allocation, 4–6
merging code, 96–98
Model-View-Controller, 172
Model-View-Presenter, 172
monitoring systems, optimizing, 105–108

Moore's law, 4
"The Morning Paper" (Colyer), 25
multicore hardware, 99–101
The Mythical Man Month (Brooks), 260, 298

N

"The Naming of Cats" (Eliot), 253
naming schemes, 1, 43–44, 252–255, 291–292
Nemeth, Evi, 297
nested files, 16–17
Network File System (NFS), 88–89, 211–214
network protocols
    streaming versus stop-and-go, 211–214
    writing, 189
        design recommendations, 196–199
        ephemeral ports, 193–195
        overallocating ports, 190–192
        port squatting, 221–223
        sequence numbers, 200–202
        standards implementation, 230–233
        TCP/IP usage versus, 224–226
Network Time Protocol (NTP), 201–202
networked systems
    debugging, 203–207
    ephemeral ports, 193–195
    latency, 208–214
    overallocating ports, 190–192
    scaling, 215–220
    security, 227–229
networking test system, 86–89
NFS (Network File System), 88–89, 211–214
nitpicking, 239–241
nonblocking I/O, 219–220
nonconcurrent systems, 99–101
NTP (Network Time Protocol), 201–202
null encryption, 91–92
nvi editor source base, 66

O

object-oriented languages
    abstractions in, 133–135
    C versus C++ 55–57
obsolescence, 263–266
opcodes for packet classification, 29
open source, 191–192, 287–289
operating systems, API changes, 141–142
optimization
    hardware and, 61–64
    of monitoring systems, 105–108
    with null encryption, 91–92
organizing log output, 116
oscilloscopes, 66

Ousterhout, John, 143, 147
overallocating ports, 190–192
overspecialization, 264–265

P

P2P (peer-to-peer) systems, 185–187
packet classification, 29
packets
    capturing, 203–207
    sequence numbers, 200–202
Parkinson, C. Northcote, 240
Parkinson's Law and Other Studies in Administration (Parkinson), 240
passwords
    PINs as, 112–114
    recovery questions, 167–168, 170
PCI Utilities, 121, 122
PDFs, documentation as, 109–111
peer-to-peer (P2P) systems, 185–187
performance optimization
    hardware and, 61–64
    of monitoring systems, 105–108
    with null encryption, 91–92
phishing, 166–171, 272
Pike, Rob, 63, 296, 297
PINs, 112–114
PKI (Public Key Infrastructure), 227–229
point-to-point networks, 225–226
polling, 105–108
ports
    ephemeral ports, 193–195
    overallocating, 190–192
    squatting, 221–223
Postel, Jon, 196
The Practice of Programming (Kernighan and Pike), 63, 296, 297
Precision Time Protocol (PTP), 201–202
presentation layer, 172–175
privacy of data, 176–180
products, systems versus, 102–104
profilers, 70
programming languages
    choosing, 45–47, 55–60
    learning, 284–286
programs, origin of term, 127–128
protocols. *See* network protocols
prototyping, 260–262
PTP (Precision Time Protocol), 201–202
Public Key Infrastructure (PKI), 227–229
Python
    abstractions in, 134
    when to use, 46

Q

quality of code, 58–60

R

radix code for address lookup, 29–30
RAM. *See* memory management
RAND licenses, 191–192
rate limiting, 271
reading, importance of, 299–301
rebooting, 115–116
recommended books, 293–298
Redis, 217
refactoring, 93–95
relevance of testing, 87
remote power control, 88
repeatability of testing, 87–88
repositories. *See* version control systems
resource management, memory allocation, 4–6
retaining configuration data during upgrading, 102–104
retaining test modes in code, 90–92
reuse of code
    abstractions, 131
    abuse of code versus, 13–15
    copying code, 39–41
    IPFW example, 28–30
robotics, AI and, 245–247
role definition, 265
root users, 267–269
run-on code, 19
Rust, 55

S

Salus, Peter H., 127
scaling networked systems, 215–220
scientific method
    communication and, 296
    debugging and, 52–54
    testing and, 86
scripted languages, abstractions in, 134–135
security
    authentication, 149–159
    buffer overflows, 198
    cross-site scripting, 160–165
    data breaches, 270–272
    input validation, 73–75, 160–165
    log output, 176–180
    networked systems, 227–229
    peer-to-peer (P2P) systems, 185–187
    phishing, 166–171
    PINs, 112–114

software updates, 279–280
system control, 267–269
tooling, 280
security audits, 69–70
security reviews, 279–283
sequence numbers, 200–202
session timeouts, 155–156
Shackleford, Rusty, 170
shell scripts, when to use, 46
Snyder, Garth, 297
Socratic method, 250–251
software architecture, broken builds and, 244
software maintenance, 93–95, 130–132
software updates, 279–280
source repositories. *See* version control systems
spaces, tabs versus, 123–125
speed of light, 209–210, 213
standards
    coding standards, 290–292
    for network protocols, 230–233
static analysis
    of code, 66
    integration with compilers, 117
    tuning, 280
stepping backward in debugging, 68
Stevens, Richard, 296
stop-and-go protocols, streaming protocols versus, 211–214
streaming protocols, stop-and-go protocols versus, 211–214
structures, clocksource.h example, 31–33
"Stupid Programmer Trick", 302–303
style. *See* coding style
substance. *See* coding substance
subtle approaches to code spelunking, 66
sudo, 267–269
system administrators, 263–266
system call tracers, 70
system control, 267–269
System V, 15
systems
    composeability, 127
    products versus, 102–104
systems design, 127–128
    abstractions, 129–139
    authentication, 149–159
    cross-site scripting, 160–165
    dependency analysis, 140–142
    design reviews, 248–251
    Java, 181–184
    log output, 176–180
    peer-to-peer (P2P) systems, 185–187
    phishing, 166–171
    prototyping, 260–262

threaded programming, 143–148
user interface design, 172–175

T

tabs, spaces versus, 123–125
TCP (Transmission Control Protocol)
    ephemeral ports, 194–195
    layering NFS over, 213–214
    number of ports, 190
    sequence numbers, 201
tcpdump, 203–207
TCP/IP Illustrated Volume 1 and 2 (Stevens), 296
TCP/IP protocol stack
    on System V, 15
    writing new protocols versus, 224–226
terminal I/O systems, 15
test interfaces, 87–88
testability of abstractions, 130
testing
    broken builds and, 242–244
    networking standards, 232
    networking test system, 86–89
    relevance and, 87
    repeatability of, 87–88
    retaining test modes in code, 90–92
    what to avoid, 83–85
theories in scientific method, 53
threaded programming
    danger of, 143–148
    debugging, 147
    libraries and, 147
    on multicore hardware, 99–101
Three Laws of Robotics, 246
ticketing systems, 273–275
TIME_WAIT state, 194–195
timeouts
    on cookies, 158–159
    session timeouts, 155–156
timestamps as sequence numbers, 201–202
tool usage, 276–278
tooling (security), 280
transaction identifiers, 144
Transmission Control Protocol. *See* TCP (Transmission Control Protocol)
Turing, Alan, 246
Turing test, 246

U

UNIX and Linux System Administration Handbook (Nemeth, et al), 297
updating
    log output, 34–36
    software, 279–280

upgrading, retaining configuration data, 102–104
user input validation, 73–75, 160–165
user interface design, 172–175
utility of abstractions, 130

V

Valgrind, 72
validating input, 73–75, 160–165
variables, global, 44
Vasa example (communication failures), 236–238
version control systems
    check-in comments, 48–50
    documentation in, 110
    finding code, 37–38
    when to merge, 96–98
version numbers for protocols, 198
VRRP (Virtual Router Redundancy Protocol), 191–192

W

Warren, Henry S. Jr., 297

watchpoints, 68
web browsers, authentication, 155–159
web forms
    input validation, 73–75, 160–165
    phishing, 166–171
weeding, 90
Whaley, Ben, 297
white lists, 75, 164
white-box security reviews, 281–282
whitespace, 123–125
wireless networks, lossy communication, 225–226
wireshark, 203–207
writing network protocols, 189
    design recommendations, 196–199
    ephemeral ports, 193–195
    overallocating ports, 190–192
    port squatting, 221–223
    sequence numbers, 200–202
    standards implementation, 230–233
    TCP/IP usage versus, 224–226

X

XSS (cross-site scripting), 160–165

Credits

Cover: Illustrations by Silky/Shutterstock, Alexslb/Shutterstock, Little Princess/Shutterstock

Page xiii: "So now I'd like to throw . . . problem. Someone bald." Kode Vicious, The Worst Idea of All Time, Development, Volume 17, issue 1, March 18, 2019.

Page 4: "Data expands to fill the space available for storage." C. Northcote Parkinson, Parkinson's Law: The Pursuit of Progress, London, John Murray, 1958.

Page 7: "No one will ever need more than 640K of RAM." Quote attributed to, but denied by Bill Gates, 1982.

Page 11: "As long as the roots . . . and summer again." Jerzy Kosinski, Being There, Corgi Books, 1970, p. 179. © 2020 Penguin Random House.

Page 13: "It can be better . . . code reuse." Rob Pike, SPLASH 12: Proceedings of the 3rd annual conference on Systems, programming, and applications: software for humanity, Association for Computing Machinery, October 2012, pp. 5–6.

Page 16: "It's turtles, all the way down." Barker, Joseph, Barker & Berg Discussion & Four Sermons, Boston: J. B. Yerrinton & Son, Printers, 1854, p. 48.

Page 18: "Make everything as simple as possible, but not simpler." Albert Einstein, Theoretical physicist.

Page 24: "Programmers are often angry because they're often scared." Paul Ford, "What is Code?", Bloomberg Businessweek, 6/11/15 © Bloomberg L.P.

Page 25: "Almost all catastrophic . . . File System) cluster." Adrian Colyer, Simple testing can prevent most critical failures, October 6, 2016.

Page 26: "This difference is likely . . . in handling these errors." Yuan et al., Simple Testing Can Prevent Most Critical Failures: An Analysis of Production Failures in Distributed Data-Intensive Systems, USENIX Association, October 6–8, 2014. © Usenix.

Page 39: "Plagiarize! Let know one . . . call it research." Tom Lehrer, Songs & More Songs, "Lobachevsky," January 22, 1953. © Tom Lehrer.

Page 42: "I've got a little list, . . . be missed." W. S. Gilbert, Arthur Sullivan, The Mikado, "I've Got a Little List".

Page 51: "The time has come . . . of many things." Lewis Carroll, Through the Looking-Glass, and What Alice Found There, Macmillan Publishers, December 27, 1871.

Page 52: "The good thing about science . . . believe in it." Neil deGrasse Tyson, American astrophysicist.

Page 55: "My name is Ozymandias . . . Mighty, and despair!" Percy Bysshe Shelley, Ozymandias, The Examiner, January 11, 1818.

Page 58: "Move fast and break things." Mark Zuckerberg, Facebook's old motto. © Mark Zuckerberg.

Page 61: "People who are . . . their own hardware." Alan Kay, American computer scientist.

Page 65: "Fools ignore complexity; . . . geniuses remove it." Alan Perlis, American computer scientist.

Page 67: "Be prepared." Lieut. Gen. Baden Powell C.B., Scouting for Boys, Horace Cox, January 24, 1908.

Page 73: "If builders built buildings . . . would destroy civilisation." Gerald M. Weinberg, American computer scientist.

Page 76: "Documentation is like . . . better than nothing." Dick Brandon, computer scientist.

Page 80: "Incorrect documentation is often worse than no documentation." Bertrand Meyer, French academic, author, and consultant in the field of computer languages.

Page 83: "Testing is the process of . . . to the anonymous." James Marcus Bach, software tester.

Page 86: "Optimism is an occupational hazard of programming: feedback is the treatment." Kent Beck, Extreme Programming Explained: Embrace Change, 2nd ed., Addison-Wesley, 2005, p. 31. © 1996–2020 Pearson.

Page 90: "It's hard enough to find . . . is error-free." Steve McConnell, Code Complete, 1st ed., Microsoft Press, 1993. © Microsoft 2020.

Page 93: "All programming is maintenance . . . writing original code." Dave Thomas, author; from Orthogonality and the DRY Principle, A Conversation with Andy Hunt and Dave Thomas, Part II, by Bill Venners (https://www.artima.com/intv/dry.html#:~:text=Dave%20Thomas%3A%20All%20 programming%20is,back%20and%20make%20a%20change.&text=But%20you%20are%20 very%20quickly,with%20a%20fresh%20source%20file), March 10, 2003.

Page 99: "Software gets slower faster than hardware gets faster." Niklaus Wirth, A Plea for Lean Software, Computer 28 (2), February 1995, pp. 64. © Copyright 2020 IEEE.

Page 100: "think digital watches are a pretty neat idea" Douglas Adams, The Hitchhiker's Guide to the Galaxy, Random House, 2007. © 2020 Penguin Random House.

Page 102: "If at first you don't succeed, call it version 1.0." Pat Rice, computer scientist.

Page 109: "If you've got . . . isn't your biggest problem." Jerry Seinfeld, American comedian.

Page 112: "If someone steals . . . are very different." Bruce Schneier, American cryptographer.

Page 117: "If you lie to the compiler, it will get its revenge." Henry Spencer, computer scientist.

Page 119: "There has never . . . history of computers." Steven Levy, Hackers: Heroes of the Computer Revolution - 25th Anniversary Edition, O'Reilly Media, Inc., 2010, p. 332. © 2020, O'Reilly Media, Inc.

Page 123: "It is allowed on . . . end of their eggs." Jonathan Swift, Gulliver's Travels, Benjamin Motte, October 28, 1726.
Page 127: "Computer Science is the study of what can be automated." Donald E. Knuth, Selected Papers on Computer Science, Cambridge University Press, 1996. © 2020 University of Cambridge.
Page 127: "1. Write programs that . . . is a universal interface." Peter H. Salus, A Quarter Century of UNIX, Addison-Wesley, 1994. © 1996–2020 Pearson.
Page 129: "The art of programming . . . effectively as possible." Dijkstra, Notes On Structured Programming (EWD249), Section 3 (On The Reliability of Mechanisms), Technische Hogeschool Eindhoven, 1970, p. 7.
Page 133: "Data dominates. . . . central to programming." Rob Pike, Notes on Programming in C, February 21, 1989.
Page 136: "A program is never . . . percent complete." Terry Baker, computer scientist.
Page 140: "Nothing is so painful to the human mind as a great and sudden change." Mary Wollstonecraft Shelley, Frankenstein: Or, The Modern Prometheus, Lackington, Hughes, Harding, Mavor, & Jones, 1818.
Page 143: "Why Threads Are A Bad Idea (for most purposes)". John Ousterhout, presented at Sun Microsystem Laboratories, 1996 USENIX Technical Conference, September 28, 1996.
Page 146: "Unsafe at any speed." Ralph Nader, Unsafe at Any Speed, Grossman, 1965.
Page 149: "Security is a state of mind." Philip T. Pease, The NSA Security Manual.
Page 155: "We should treat . . . as nuclear waste". theguardian.com, January 15, 2008. © 2020 Guardian News & Media Limited.
Page 160: "There is no such . . . varying levels of insecurity." Salman Rushdie, novelist.
Page 166: "Using encryption on . . . living on a park bench." As presented in the first edition of Web Security & Commerce (O'Reilly, 1997, S. Garfinkel & G. Spafford), and originally stated in abbreviated form by Spafford in a presentation at Supercomputing 1995 in San Diego.
Page 167: "I want to impress . . . personnel through carelessness." Fictional character General Jack D. Ripper in Stanley Kubrick, Dr. Strangelove, Or: How I Learned To Stop Worrying And Love The Bomb, Hawk Films, 1964. © Hawk Films.
Page 172: "A common mistake . . . of complete fools." Douglas Adams, Mostly Harmless, Harmony Books, 1992. © 2020 Penguin Random House Company.
Page 174: "Layers are a good . . . to implement them." Van Jacobsen, networking researcher.
Page 176: "The most secure code in the world is code which is never written." Colin Percival, mathematician.
Page 181: "If Java had true . . . upon execution." Robert Sewell, computer scientist.
Page 189: "Distributed systems are hard." Jonathan Anderson, computer scientist.
Page 190: "She doesn't think she waltzes but would rather like to try." W. S. Gilbert & Arthur Sullivan, The Mikado, "I've Got a Little List".
Page 194: "*ephemeral* 'lasting for a very short time'". Dictionary.com, LLC © 2020 Dictionary.com, LLC.
Page 196: "An implementation should . . . its receiving behavior." Jon Postel, Internet Protocol, RFC 791, USC/Information Sciences Institute, September 1981. © 2020 University of Southern California.
Page 208: "When it absolutely, positively, has to be there overnight". Courtesy of FedEx Corporation. © FedEx 1995-2020.
Page 211: "Never underestimate the . . . the highway". Andrew S. Tanenbaum, Computer Networks, Pearson Education, Inc., 1981. © 1996–2020 Pearson.
Page 221: "Bureaucracies are designed . . . as its enemy." Brooks Atkinson, American critic.
Page 221: "rough consensus and running code" The Tao of IETF, A Novice's Guide to the Internet Engineering Task Force. © 2019 IETF Trust. All rights reserved.
Page 221: "The tree of liberty must occasionally be watered with the blood of patriots." Thomas Jefferson, The "Tree of Liberty" letter, November 13, 1787.
Page 224: "Everything is derivative. . . . of giants." Brendan Byrnes, An Interview with Reddit Co-Founder Alexis Ohanian, fool.com, October 17, 2013.
Page 225: "If only we could . . . and ours is shiny and new." George Neville-Neil: Bugs and Bragging Rights, in: Queue 11.10 (Oct. 2013), 10–12, URL: https://doi.org/10.1145/2542661.2542663. © Association for Computing Machinery.
Page 227: "Relying on the . . . your window blinds." John Perry Barlow, American poet and essayist.
Page 230: "The nice thing . . . next year's model." Rear Adm. Grace Murray Hopper, American computer scientist.
Page 235: "No matter what the problem is, it's always a people problem." Gerald M. Weinberg, The Secrets of Consulting, Dorset House Pub., 1985. © 1996-2008 by Dorset House Publishing Co., Inc.
Page 236: "Pride goeth before destruction, and a haughty spirit before a fall." King James Bible, Proverbs, 16:18.
Page 239: "Why should I care what color the bikeshed is?" Poul-Henning Kamp, Danish software developer.
Page 242: "I don't care if . . . shipping your machine!" Vidiu Platon, computer scientist.
Page 245: "A robot may not . . . come to harm". Isaac Asimov's, Runaround - First of Three Laws of Robotics, 1942. © Isaac Asimov.

Page 247: "Any sufficiently advanced science is indistinguishable from magic." Arthur C. Clarke, The Sentinel: Masterworks of Science Fiction and Fantasy, Barnes & Nobles Books by arrangement with Byron Preiss Visual Publications, 1996. © Arthur C. Clarke.

Page 248: "Perfection is achieved, ... left to take away." Atoine de Saint Exupery, Airman's Odyssey, Harcourt, Brace & World, 1942. © 2020 Houghton Mifflin Harcourt.

Page 254: "M-I crooked letter ... hump back I". Dan MacIntosh, Mac Powell of Third Day quoted in Mac Powell, Mississippi, 2012.

Page 260: "Nine mothers cannot make a baby in one month." Frederick P. Brooks, Jr. The Mythical Man-Month, Addison-Wesley, 1975. © 1996–2020 Pearson.

Page 263: "Perhaps it is ... they make. Or sell." Studs Terkel, American author.

Page 267: "The possession of great power necessarily implies great responsibility." William Lamb, British M.P, 1817.

Page 270: "We don't care. We don't have to. We're the phone company." Character of Ernestine, created by Lily Tomlin, American comedian and actress.

Page 271: "The self-described hackers ... phishing at www.att.com/safety." Dorothy Attwood, Senior Vice President, Public Policy and Chief Privacy Officer for AT&T.

Page 273: "In the beginning was the word and the word was [bulls#!+]." William S. Burroughs, The Ticket That Exploded, Olympia Press, 1962. © Grove/Atlantic.

Page 276: "There's nothing more permanent than a temporary hack." Kyle Simpson, Evangelist of the open web.

Page 281: "If architects designed buildings ... all of society." Gerald M. Weinberg, Weinberg's Second Law.

Page 287: "The GNU GPL was not designed to be open source." Richard M. Stallman, Re: GPL version 4" on NetBSD mailing list, mail-index.netbsd.org. July 17, 2008.

Page 290: "Turn him to any ... his garter". Shakespeare, William, Henry V, Act 1 Scene 1, 45–47.

Page 293: "You don't have to ... stop reading them." Ray Bradbury, Fahrenheit 451, Ballantine Books, 1953. © 2020 Penguin Random House.

Page 293: "ninety percent of everything is crap". Theodore Sturgeon, Sturgeon's law.

Page 297: "Time enough at last!" Henry Bemis in The Twilight Zone: Season 1, Episode 8, November 20, 1959. © Cayuga Productions, Inc.

Page 299: "Time isn't the main thing. It's the only thing." Miles Davis, American jazz musician.

Page 302: "Oh Bullwinkle, that trick never works!" Character of Rocky J. Squirrel in Rocky and Bullwinkle Show, produced by Jay Ward and Bill Scott. © Jay Ward and Bill Scott.

The Craft of Programming Books & eBooks

Be a Better Programmer

InformIT has the most recommended programming books of all time. Check out our guides to help you write better code.

Visit **informit.com/programming/craft** to shop and preview sample chapters from classic guides on programming including

> *The Pragmatic Programmer*
> *The Art of Computer Programming*
> *Clean Code*
> *Refactoring*
> *Code Complete*
> *Working Effectively with Legacy Code*
> *Hacker's Delight*
> *Programming Pearls*

Addison-Wesley · Adobe Press · Cisco Press · Microsoft Press
Pearson IT Certification · Que · Sams · Peachpit Press

Photo by izusek/gettyimages

Register Your Product at informit.com/register

Access additional benefits and **save 35%** on your next purchase

- Automatically receive a coupon for 35% off your next purchase, valid for 30 days. Look for your code in your InformIT cart or the Manage Codes section of your account page.
- Download available product updates.
- Access bonus material if available.*
- Check the box to hear from us and receive exclusive offers on new editions and related products.

*Registration benefits vary by product. Benefits will be listed on your account page under Registered Products.

InformIT.com—The Trusted Technology Learning Source

InformIT is the online home of information technology brands at Pearson, the world's foremost education company. At InformIT.com, you can:

- Shop our books, eBooks, software, and video training
- Take advantage of our special offers and promotions
- Sign up to receive special offers and monthly newsletter
- Access thousands of free chapters and video lessons

Connect with InformIT—Visit informit.com/community

the trusted technology learning source

Addison-Wesley • Adobe Press • Cisco Press • Microsoft Press • Pearson IT Certification • Que • Sams • Peachpit Press